T0265519

THE LANGUAGE OF CLIMATE POLITICS

THE LANGUAGE OF CLIMATE POLITICS

FOSSIL-FUEL PROPAGANDA AND HOW TO FIGHT IT

GENEVIEVE GUENTHER

OXFORD
UNIVERSITY PRESS

Oxford University Press is a department of the University of Oxford. It furthers the University's objective of excellence in research, scholarship, and education by publishing worldwide. Oxford is a registered trade mark of Oxford University Press in the UK and certain other countries.

Published in the United States of America by Oxford University Press
198 Madison Avenue, New York, NY 10016, United States of America.

Library of Congress Cataloging-in-Publication Data
Names: Guenther, Genevieve Juliette, author.
Title: The language of climate politics : fossil-fuel propaganda and how to fight it / Genevieve Guenther.
Description: New York, NY : Oxford University Press, 2024. |
Includes bibliographical references and index. |
Identifiers: LCCN 2024002565 (print) | LCCN 2024002566 (ebook) |
ISBN 9780197642238 (hardback) | ISBN 9780197642252 (epub) |
ISBN 9780197642269
Subjects: LCSH: Communication in climatology. | Climatic changes—Political aspects. |
Fossil fuels—Environmental aspects. | Petroleum industry and trade—Environmental aspects. | Environmental sciences—Language.
Classification: LCC QC902.9 .G84 2024 (print) | LCC QC902.9 (ebook) |
DDC 363.7001/4—dc23/eng/20240222
LC record available at https://lccn.loc.gov/2024002565
LC ebook record available at https://lccn.loc.gov/2024002566

DOI: 10.1093/oso/9780197642238.001.0001

Printed by Sheridan Books, Inc., United States of America

for Teddy

The problem in democratic societies lies in figuring out which apparently nonpropagandistic claims are in fact propaganda.

—Jason Stanley, *How Propaganda Works*

To clarify thought, to discredit the intrinsically meaningless words, and to define the use of others by precise analysis—to do this, strange though it may appear, might be a way of saving human lives.

—Simone Weil, "The Power of Words"

Now is the time to speak clearly.

—Greta Thunberg, *No One Is Too Small to Make a Difference*

Contents

Preface: We

It feels perfectly natural to use the word "we" when you talk about climate change. "*We* are causing climate change." "*We* are emitting more carbon dioxide than ever." "*We* need to draw emissions down to net zero in order to halt global heating at the Paris Agreement target of well below 2° Celsius."

Given that human beings are in fact causing climate change, the impulse to use the word "we" makes sense. But there's a real problem with it: the guilty collective it invokes simply doesn't exist. The "we" responsible for climate change is a fictional construct, one that's distorting and dangerous. By hiding who's really responsible for the crisis, the word "we" provides political cover for the people who are happy to destroy a livable climate to gain more profit and power.

Let's think about it. Who is this "we"? Does it include the nearly 700 million people who live on less than $2.15 a day?[1] Does it include the approximately 4.2 billion people, half the global population, who live on less than $6.85 a day?[2] Does it include the millions of people all over the world—like the six million who participated in the global climate strikes of September 2019—doing whatever they can to lower their own emissions and push for systemic change? Does it include Bill McKibben, the elder statesman of the climate movement who wrote his first book about climate change in 1989? How about Greta Thunberg, the young woman who inspired worldwide climate protests by sitting in front of the Swedish Parliament every Friday to demand her government take action at the scale of the crisis? Does it include the indigenous peoples who have been living in harmony with their ecosystems for generations?[3] Does it include our children?

Of course the universal "we" seems real. The fossil-fuel system, for the moment, feels all-encompassing. It provides the means for what people do on this planet. In its inclusions and exclusions, its laying out the conditions for human action, it seems totalizing, especially from an affluent American vantage point. But it's not totalizing. And it's certainly not eternal. It requires active reproduction at every moment in time: through subsidies, through construction and repair of its infrastructure, through court cases that uphold its laws, through protection of its "assets" by the military, through Instagram photos that pretend its benefits will bring you joy, and on and on.

Instead of thinking of climate change as something "we" are doing, always remember that there are millions, possibly billions, of people on this planet who would rather preserve civilization than destroy it with climate breakdown, who would rather have the fossil-fuel economy end than continue. Those people are not all mobilized, by any means, but they are there.

But remember too that there are millions of other people, some of them running the world, who seem willing to destroy civilization and allow untold numbers of people to die in the decades ahead so that the fossil-fuel system can continue now.

Remember as well that there are degrees of complicity. Without structural changes paid for collectively, most of us have no alternative but to use fossil fuels to some extent. Each of us can only do our best. And lots of people—including, as you shall see, some climate researchers, policymakers, and even advocates—believe, mistakenly, that the world can keep using coal, oil, and methane gas and still halt global heating anyway. But neither constrained choices nor mistaken beliefs are akin to the deep and shameful complicity of, for example, those in public relations who create advertising campaigns for oil and gas companies or those in the news media who refuse to mention climate change or the dangers of fossil fuels in their reporting. Such people are making money destroying the world.

Complicit people and institutions must be called out and encouraged to change. And the fossil-fuel industry must be fought, and the governments that support that industry must be replaced. But none of us will be effective in this if we think of climate change as something "we" are doing. To think of climate change as something that "we" are doing, instead of something we are being prevented from undoing, perpetuates the very ideology of the fossil-fuel economy we're trying to transform.

Fossil-fuel ideology can be reproduced even by a tiny, innocuous pronoun like "we." This capacity to transmit ideologies—to shape the way people see the world, without their even being aware—makes words politically powerful. To undo climate change, a new collective "we"—me, you, everyone who reads this book, everyone with whom we share its ideas—will need to use the power of words to fight climate propaganda and transform the deep ideologies of the fossil-fuel economy. Contributing to that transformation is the goal of this book.

Introduction: Why Words Matter in Climate Politics

> Political ground is gained not when you successfully inhabit the middle ground, but when you successfully impose your framing as the "common-sense" position.
>
> —George Lakoff, University of California, Berkeley ("Conservatives Don't Follow the Polls, They Want to Change Them")

Climate change is "a hoax," Donald Trump sneers, using his best wise-guy accent. This is the kind of lie that most people associate with fossil-fuel propaganda. And for good reason. Since at least the 1970s, coal, oil, and gas companies have known that their products would cause the planet to heat up, undermining the climate that enabled civilization to flourish over the past 10,000 years; to cover up what they knew, these companies worked with political strategists, advocacy organizations, think tanks, trade associations, and advertising firms to hone and spin the message that climate change wasn't real.[1] By the time the US Congress first attempted to take major climate action, with a 2009 cap-and-trade emissions bill that died in the Senate, this fossil-fuel propaganda had become the Republican party line. It remains the party line—at least for the MAGA base. In the first Republican primary debate of the 2024 election, entrepreneurial candidate Vivek Ramaswamy, perhaps gunning to become Trump's vice-presidential pick, declared that "the climate change agenda is a hoax!"

"The reality is," Ramaswamy added, "more people are dying of bad climate change policies than they are of actual climate change."[2] For this latter claim he offered no evidence, of course.

As absurd as it is to say that climate-change policies are killing people, it is equally absurd, at this point, to claim that climate change isn't real. The planet is over 1.3°C hotter on average than it was in the pre-industrial era. Its temperature is rising at a rate of around 0.2° per decade; land maximum temperatures are rising twice as fast, to over 1.7° already.[3] Every summer, a ghastly parade of stories about deadly extreme weather marches across the news, and scientists are now able to attribute this weather to climate change. In 2015, over 195 nations signed the Paris Agreement, pledging to halt global warming at 1.5° or, at worst, "well below 2 degrees." By the end of 2022, 79 percent of global emissions were covered by net-zero targets.[4] (Net zero is the point where emissions are virtually eliminated, and any unavoidable climate pollution left over is taken out of the atmosphere by carbon removal.) Trump and his acolytes are perfectly willing to dismiss all this, telling people to reject the evidence of their own eyes and ears. But other fossil-fuel interests—Republicans without Trump's authoritarian charisma, titans of finance and tech, fossil-energy-funded researchers at elite universities, conservative media magnates, and, strikingly, coal, oil, and gas executives themselves—have started spreading a new, more subtle form of propaganda, which acknowledges that climate change is real but still seeks to justify continuing the fossil-fuel economy.

This propaganda is spun out of six key terms that dominate the language of climate politics: *alarmist*, *cost*, *growth*, *"India and China,"* *innovation*, and *resilience*. Together these terms weave a narrative that goes something like this: "Yes, climate change is real, but calling it an existential threat is just *alarmist*—and anyway phasing out coal, oil, and gas would *cost* us too much. Human flourishing relies on the economic *growth* enabled by fossil fuels, so we need to keep using them and deal with climate change by fostering technological *innovation* and increasing our *resilience*. Besides, America should not act unilaterally on the climate crisis while emissions are rising in *India and China*." This narrative is designed to foment the incorrect and dangerous belief that the world does not need essentially to stop using fossil fuels—either because climate change won't be that destructive or, in some versions of the story, because the world can keep using coal, oil, and gas and still halt global heating anyway.

It would be bad enough if this narrative was being repeated only on the right. But what gives this narrative its power over current climate politics is that it's repeated, in echoes and fragments and sometimes in its entirety, by people on both sides of the supposedly partisan climate-change divide—not just by fossil energy interests, but also by scientists, economists, journalists, politicians, and sometimes even activists, all of whom sincerely intend to advance climate solutions. But that's because, as we shall see, these terms and the ideas they convey are appropriated from language produced over the decades by these groups themselves. Fossil-energy interests have mined the language of climate advocates for material they can use for propaganda purposes, extracting, twisting, and deploying their words to entrap those advocates into unwittingly normalizing fossil-fuel disinformation. This dynamic turns fossil-fuel propaganda into a bipartisan consensus—the common-sense position.

We can see how this dynamic plays out by looking at the word "uncertainty." As the Harvard historian of science Naomi Oreskes showed in *Merchants of Doubt*, fossil-fuel partisans created doubt about the reality of climate change by talking incessantly about scientific "uncertainty" in their public statements about global heating. And for decades this strategy worked. They were able to use the word "uncertainty" so effectively in part because they exploited the word's cultural ambiguity—a quality that arises when a word means one thing to a group of people in a particular subculture but another thing to the general public. Most people understand "uncertainty" to mean something like "not knowing for sure" or "not having concluded yet." Scientists sometimes use the word in this way too, of course, but not always. When climate scientists speak of the "uncertainty" of their findings, they usually refer to the range of possible outcomes they can project with confidence. Indeed, "uncertainty" and "confidence" are actually synonyms in scientific discourse: scientists can say either that a model produces an "uncertainty interval" or that it produces a "confidence interval." And those intervals do not indicate "either it happens or it doesn't." Rather they span everything that could possibly happen—from the best to the worst-case outcomes. So when scientists were being the most scrupulous in their communications about the climate crisis, trying to give the public a full sense of what they confidently knew at the time, they unwittingly echoed the message that there was some "uncertainty" that meant they weren't sure if climate change was real. In a truly fiendish act of appropriation, fossil-fuel interests managed to recruit scientists into inadvertently spreading doubt.

Although oil and gas interests no longer produce disinformation by manipulating the ambiguities of the word "uncertainty," they still use this technique of linguistic appropriation to shape the new, twenty-first-century propaganda that says, falsely, that phasing out fossil fuels would be worse than climate change itself. But let's be clear. Both the International Energy Agency and the United Nations Intergovernmental Panel on Climate Change (the IPCC) have stated unequivocally that in order to achieve net-zero emissions in time to halt global heating at a relatively safe level, the development of new oil and gas fields must cease immediately. Both have also stated, unequivocally, that to have even a 50–50 chance to halt global heating at the Paris target of 1.5°, *existing* fossil-energy infrastructure must be retired before the end of its expected lifetime.[5] And fossil-energy producers know this. A 2023 "Energy Security Scenario" released by Shell, for example, saw the immediate end to growth in oil and gas production in models that kept heating below 1.5°C.[6]

Yet the American right wing continues its unequivocal support for expanding coal, oil, and methane gas. "This isn't that complicated, guys," Ramaswamy said at the 2024 Primary debate, "unlock American energy, drill, frack, burn coal."[7] And Republicans do not just support fossil energy; as we shall see throughout this book, they also consistently obstruct or, if need be, reverse any legislative, executive, private-sector, or cultural support for climate-safe energy. In the summer of 2023, then the hottest summer in human history, over 400 conservative analysts and former Trump administration officials organized by the Heritage Foundation, the think tank that has shaped GOP policy since the Reagan administration, released a 920-page plan to dismantle most of the federal government's capacity to help fight climate change. Called "Project 2025," the plan details how officials could shutter the Energy Department's renewable energy offices; prevent states from adopting California's car pollution standards; block the expansion of the electrical grid for wind and solar energy; and task Republican state officers with the oversight, such as it would be, of the fossil-fuel industry. The timeline in the plan begins on day one of a Republican presidency.[8]

The Democrats have a much more ambivalent relationship with fossil fuels. In his first term, President Biden was perhaps the best climate advocate the White House had ever seen, and with the slimmest of Democratic majorities in the Senate he managed to shepherd through two major pieces of climate legislation. The 2021 Infrastructure Investment and Jobs Act invested in transmission and power-grid upgrades and provided grants for electric-vehicle infrastructure deployment as well as the research and development of nascent technologies like direct air capture and green hydrogen. And the 2022 Inflation Reduction Act made major investments in utility-scale and distributed solar, wind, and other renewable resources—along with carbon-capture facilities in the power and industrial sectors—and it provided uncapped tax credits for the purchase of electric vehicles, heat pumps, and other clean-energy consumer products along with more support for direct air capture and green hydrogen.

At the same time, however, in the first two years of Biden's presidency, his administration approved more permits for oil and gas drilling on public lands than had even his climate-denying predecessor.[9] The United States was already the world's biggest oil and gas producer, and one of the biggest fossil-fuel exporters.[10] Yet in 2021, US methane gas production increased by 2 percent; in mid-2022 the US Energy Information Administration (EIA) projected production would increase by 3 percent that year and an additional 2 percent in 2023. The EIA expected that crude oil production would increase by 6 percent in 2022 and 5 percent in 2023.[11] As of January 2024, the EIA saw US methane gas and crude oil production rising through 2025 to "new records."[12] Major emitters are supposed to be halting new projects and phasing out existing fossil-fuel infrastructure. Even under Biden, the United States has been doing the opposite. One analysis found that the United States is poised to be the world's largest expander of oil and gas extraction between 2023 and 2050.[13] What this means, realistically, is that if they are not changed, America's current climate politics, even under a Democratic President leading a Democratic Congress, will lead to the planet heating up beyond the Paris targets. In 2022 the United Nations Environment Programme found that the legislative and executive actions already in place across the world would likely lead to 2.8°C of heating by 2100, with a confidence interval of 1.9–3.3°C.[14] Scientists have called 3° of warming "a catastrophic outcome."[15]

The year 2100 may seem like a long time away, but it isn't. My own son was born in 2010. His life will play out across this century, when the world will either halt global heating at a manageable level or unravel. All this is no longer about "future generations," but the families we have in our homes today. As the IPCC said in its 2023 report: "Any further delay in concerted global action will miss a brief and rapidly closing window to secure a livable future."[16]

To secure a livable future, one thing we will need to do is dismantle and reframe the terms dominating the language of climate politics. This is a key element of the political work to do now. Vast majorities of Americans support getting emissions to net zero: a large 2022 Pew research poll found that 69 percent of Americans favor the United States becoming carbon neutral by 2050.[17] But only a fraction of Americans support phasing out fossil fuels entirely. A mere

31 percent say that the United States should completely phase out oil, coal, and methane gas, according to a 2023 poll, while 68 percent say the country should use a mix of energy sources, including fossil fuels. These numbers are skewed by partisanship, of course, since only 12 percent of Republican voters support a fossil-fuel phaseout, but still—only a minority of Democrats, 48 percent, support running the economy on 100 percent climate-safe energy.[18]

Fossil-energy propaganda profoundly influences the political views of this muddled middle and the elites who represent them. I'll show how it does so by telling the stories of the main players in climate politics: the world-renowned climate scientists intimidated into downplaying the dangers of global heating by decades of relentless attacks on their reputations; the ivy-league economists celebrated by both the Nobel Prize committee and fossil-fuel front groups; the carbon-management lobbyists appointed to the Department of Energy by the Biden administration; and the democratic and authoritarian leaders colluding behind closed doors at high-level international climate negotiations. What these people say about fossil fuels and the climate crisis circulates through many platforms, high and low: scientific reports, policy briefings, political speeches, press-conference transcripts, economics papers, academic and popular books about climate change, court cases, internal corporate and interest-group memos, oil and gas advertisements, debate transcripts, social media posts, climate-negotiation decision texts, and, not least, news-media reporting. By following the key words of climate politics across these platforms, you will learn to recognize the mechanisms behind oil and gas interests' newest propaganda. You will see how staunch opponents and ambivalent advocates of climate action reinforce each other's messages and reaffirm fossil-fuel ideologies. And through this analysis you will be introduced to new ways to talk about the climate crisis—message-tested in focus-group polls commissioned by my organization End Climate Silence. This new language will be hard for fossil-fuel interests to appropriate and, crucially, it will keep the phaseout of fossil fuels at the center of the climate-change agenda.

Of course, talking alone won't halt global warming. But, as they say on the right, ideas have consequences. Insofar as words shape ideas and ideas influence actions, we will need to transform the language of climate politics just as we need to transform the more material ways our politics supports the fossil-fuel economy. We will need to speak with the rhetorical strategies of our opponents in mind so we can stop echoing their language and normalizing their propaganda. For there is no reason to normalize the false beliefs of the fossil-fuel economy, to proceed as if the world couldn't be otherwise. The world *can* be otherwise. What follows will enable us to start talking about that new world—and help bring it into being.

I

Alarmist

Warming substantially larger than the assessed very likely range of future warming cannot be ruled out.

—The United Nations Intergovernmental Panel on Climate Change (*Climate Change 2021: The Physical Science Basis*)

I have been surprised and alarmed at the record temperatures and floods we have seen in many places around the world with only 1.1°C warming.

—Sir Brian Hoskins, Imperial College London, The Royal Society ("Why Scientists Are Using the Word Scary over the Climate Crisis")

[Climatic] changes would, however, most impact humans . . . Civilization could prove a fragile thing.

—Shell PLC ("Confidential Group Planning Scenarios, PL89 SO1, October 1989")

A favorite propaganda tactic of fossil-fuel interests is to call people who speak out about climate change "alarmists." To call someone an alarmist is to accuse them of being an unreliable narrator—a person who cannot be trusted to tell the truth about global heating, either because they are emotionally unstable or, conversely, because they cunningly exaggerate the dangers of heating in order to frighten people into accepting their preferred policies. Fossil-energy spokespeople use this tactic not only against climate activists trying to build public support for clean energy, but also against scientists simply reporting the results of their research. So, for example, analysts at the American Enterprise Institute—a conservative think tank whose funders have included ExxonMobil, Koch foundations, and the Bill and Melinda Gates Foundation—have argued that there is "no evidence that a climate 'crisis' looms in our future" and that the reports of the United Nations Intergovernmental Panel on Climate Change (IPCC) are nothing more than "unfounded alarmism."[1] Mocking "climate alarmists" for supposedly warning that "the world only has one year, two years, ten years and so on to substantially reduce fossil-fuel consumption, or else it will be too late to prevent multitudes of cascading global disasters that make the Biblical plagues of Egypt pale by comparison," the director of the Arthur B. Robinson Center on Climate and Environmental Policy at the Heartland Institute, another prominent right-wing think tank, dismisses the need for decarbonization entirely, calling scientific research "climate delusion" whose purpose is nothing more than "ever more intrusive control" over peoples' lives.[2] Repeating this trope, the influential Fox News talk-show host Sean Hannity regularly characterizes climate scientists as "climate cult alarmists" who have fabricated global heating as a "huge Trojan horse for big-government socialism."[3]

It may seem as if the accusation that climate scientists are alarmist cult leaders would have nothing to do with reality-based climate discourse—certainly not with the public speech of climate advocates or climate scientists themselves. But in the language of climate politics, no neat partisan divide separates the people accusing advocates and scientists of alarmism and those who stand accused. Rather, the discourse of alarmism functions like a network connecting disparate ideological groups. Those on the right who would deny that climate change is real are connected to centrist techno-optimists who acknowledge the reality of climate change but deny its dangers. Many of these techno-optimists criticize left-wing scientists and advocates who warn of catastrophe if the world fails to phase out fossil fuels, and some of these more alarmed advocates can sound like the so-called "doomers"—those who wrongly believe that the climate has already passed its tipping points or that civil society will never build enough power to depose fossil-energy incumbents. The composition of these groups can also be unexpected: sometimes self-professed "environmentalists" deny that climate change threatens the stability of our civilization, and sometimes cool-headed mainstream journalists write with the voice of doom. Commentators can also change their positions in this network from one year to the next—and some can occupy more than one position at any given time—making the different categories of climate discourse fluid and shifting. Still, mapping out the dynamics of the network of climate alarm will shed light on the key players' positions on fossil fuels and the political dynamics underlying the climate crisis itself.

These dynamics have played out through the rise and fall of a recent narrative about how people should understand the climate future. From roughly 2018 to the beginning of the Covid-19 pandemic, climate alarm came to dominate the public conversation about the climate crisis, and this alarm helped to motivate a global climate movement. As one might expect, fear of climate change was attacked by right-wing spokespeople for the fossil-energy industry, but, as one might not expect, it was also criticized by techno-optimists and sometimes climate scientists themselves. These scientists were amplified by journalists who covered the political shifts enforced by civil society as an uplifting story of market forces creating progress. What emerged out of this period was a reassuring promise that the world had averted the catastrophic future that scientists used to fear and there was less need for alarm. But this message turned out to be premature. With truly staggering speed, the planet has revealed that the climate system will likely break down in catastrophic ways at much lower levels of global warming than once predicted. So even though scientists have lowered their estimates of future emissions, the world is in no less danger than it ever was. This terrible truth, I will show, requires a new message: it is appropriate to be scared by what will happen if we—alarmed citizens everywhere—do not force elected officials to stop supporting fossil fuels, yet the best way to ensure a safer future is not to dwell on the impacts of climate change, but to turn our attention to the people who are doing everything they can to keep the fossil energy system in place. We must focus on removing those people from cultural and political power.

Lukewarmers

Even if some major right-wing media outlets are still trying to pretend that climate change isn't real, many Republican politicians have started distancing themselves from this strategy, perhaps recognizing how increasingly absurd it appears. Instead, they circulate the more subtle propaganda that they accept the reality of the climate crisis but view warnings about its dangers as "alarmism." Wisconsin Senator Ron Johnson, for instance, introduced this talking point during a 2023 Budget Committee Hearing: "I'm not a climate denier," he waffled, "I'm just not a climate alarmist."[4] Johnson here performs a kind of half-climate-denial that Robert Ward, a communications director at the London School of Economics, has called "lukewarmism."[5] Lukewarmers ignore or misrepresent the scientific evidence that global warming over 2°C will be catastrophic for humanity, while also insisting, falsely, that a non-fossil energy system built with current technologies would create miserable poverty, in order to make people feel that the benefits of continuing to use fossil fuels would far outweigh the dangers of even high levels of warming.

One of the most prominent lukewarmers is Bjørn Lomborg, the founder of the philanthropy advisory The Copenhagen Consensus. Lomborg regularly appears on Fox News and publishes editorials in the *Wall Street Journal* arguing that climate change is mostly benign. In his 2020 book, *False Alarm*, he notes that "global warming is real," but claims that it's "a manageable problem," whose "overall negative impact on the world . . . will pale in comparison to all of the positive gains" from using fossil fuels "we have seen so far and will continue to see in the century ahead."[6] Every time Lomborg publishes a version of this argument, climate scientists and analysts patiently illustrate how he cherry-picks and falsifies data to support it—indeed, whole books have been written on the subject, such as the aptly titled *The Lomborg Deception*.[7] Here we will turn to just one of his falsehoods, namely his claim about Americans' climate views in *False Alarm*, where he cites a British YouGov poll as showing that in the United States "four of ten people believe global warming will lead to mankind's extinction."[8] But this poll says no such thing. What it says is that Americans are "far less likely" than people in other regions "to think climate change will cause . . . human extinction"—indeed, that only 10 percent of Americans believe that "it's already too late to avoid the worst effects of climate change."[9] It is on misrepresentations like these that Lomborg's arguments are built.[10]

Perhaps surprisingly, some lukewarmers are self-professed environmentalists. Take, for instance, the analysts at the ecomodernist California think tank The Breakthrough Institute (BTI). The BTI is well respected by some thought-leaders and the mainstream press. Their 2022 conference on "Progress Problems" featured the *New York Times*' Ezra Klein, researchers from Johns Hopkins and Harvard, among other universities, and climate journalists from the *Washington Post*, *Vox*, and *The Economist*.[11] BTI analysts advocate for the expansion of nuclear, rather than solar or wind power, and for increased fossil-fuel production, particularly in the Global South where, they argue, energy poverty is a more pressing problem than climate change. In their mission statement "The Ecomodernist Manifesto," they argue that fossil fuels are the salvation of the poor: "a new coal-fired power station in Bangladesh may bring air pollution and rising carbon dioxide emissions but will also save lives."[12] This is a highly misleading statement, which cherry-picks data in prime Lomborgian fashion, ignoring that air pollution from fossil fuels kills millions of people each year, primarily in poorer nations.[13] The "Manifesto" also elides the more basic point that it is not coal but electricity that saves lives, and electricity does not need to be generated by coal-fired power plants. Nor should it be, for that matter, due to the urgent problem of "rising carbon dioxide emissions." But the people at the BTI are not concerned about those. Although they acknowledge that climate change poses the "risk of catastrophic impacts on societies," they also assure us, as does Lomborg, that "even as human environmental impacts continue to grow in the aggregate, a range of long-term trends are today driving significant decoupling of human well-being from environmental impacts."[14]

This is the lukewarmer message par excellence: climate change is not a threat to human survival, nor even to the stability of civilization; rather it is a manageable problem for which a legitimate solution is to continue to use fossil energy indefinitely for ongoing development which, supposedly, allows for the "decoupling of human well-being from environmental impacts." Indeed, BTI founder Ted Nordhaus has said that the "agenda" to phase out fossil energy is "impossible," but no one should feel dismayed about this putative fact because climate change is simply "a chronic problem that human societies will need to manage"—as if global heating were a case of planetary diabetes, rather than a cancer that needs to be caught and cured early before it metastasizes out of control.[15] Never mind that pre-eminent climate scientists such as Joëlle Gergis, a lead author of the IPCC's Sixth Assessment report, liken climate change to a cancer: "to use a medical analogy," Gergis has written, "if we choose to treat our cancer early enough we can stop its spread, but the longer we delay the more limited our treatment options and the worse the damage becomes."[16]

Techno-optimists

Positioned close to the lukewarmers, the "techno-optimists" occupy the next node in the discursive network of climate alarm. Unlike Lomborg and Nordhaus, these commentators agree that emissions must be zeroed out as quickly as possible, but they also believe that even after the world creates a global net-zero economy, a relatively large amount of ongoing fossil-fuel use will be enabled by technologies like carbon capture and solar-geoengineering. More broadly, they also believe that current systems of production and consumption should continue even in a decarbonized world. They argue that innovation will continue to improve lives, as it did in the Global North over the twentieth century, and market forces will lead us into a new, clean-energy era as technologies like solar and wind become increasingly competitive with fossil fuels. In order to encourage the transition into this era, techno-optimists say, climate communicators should avoid talking about the dangers of global heating, lest they inspire any fear or pessimism in their listeners. Instead they should keep public attention on past and future progress in order to foment the optimism that leads to a general "can-do" spirit and, more specifically, to private investment in new industries.

Yet this apolitical, purely economic understanding of the climate crisis sometimes presents a distorted view of the future and can attribute progress to market forces when it was in fact hard-won by government policy. These problems appear sometimes, for instance, in the work of the data scientist Hannah Ritchie, the author of *Not the End of the World* and the Deputy Editor at Our World in Data, an influential website that aggregates and visualizes information on public health, climate change, and the economy. In an article for WIRED magazine entitled "Stop Telling Kids They'll Die from Climate Change," Ritchie takes the climate movement to task for creating what she calls a "doomsday mindset." Ritchie suggests in this piece that climate change, although it comes with "risks," will not be all that destructive or dangerous. Criticizing news reports of increasing extreme weather disasters as creating a "false perspective" on the threats of global heating, Ritchie claims that "the data" tells a different story: "death rates from disasters have fallen a lot over the past century," she assures us, due to "better technologies to predict storms, wildfires, and floods; infrastructure to protect ourselves; and networks to cooperate and recover when a disaster does strike."

Now, whether this increased resilience will continue to be robust as the planet slips out of the climate niche in which all civilization was built, in this article Ritchie does not wonder. Instead, she professes faith in the invisible hand. Pointing to the remarkable drops in the cost of solar and wind power since 2009, Ritchie claims that these falling prices show that "we can achieve . . . progress without real political or financial support" or even despite the fact that "low-carbon technologies have been up against lobbying fossil fuel giants." She even goes so far as to imply that civil society doesn't need to worry about pushing governments to pass or implement climate policy. "Politicians might be slow," she continues "but technological change is not." Even those "who don't care about climate change" will make the consumer choice to substitute clean energy for fossil fuels, she assures us, "because it makes economic sense to do so."[17]

What this argument misses is the historical relationship between technological innovation and government policy. It is not the case that the drop in the price of wind and solar shows that "we" can achieve progress "without real political or financial support." In truth, the price of solar and wind has dropped so precipitously because starting in 2009, China classified clean energy as a "strategic emerging industry" and, passing state policy to nurture this industry, it massively subsidized not only the research and development, but the deployment, scaling, and exporting of renewable technologies.[18] The sudden and amazing competitiveness of clean energy generation *was* the product of government climate policy, just not climate policy in the West.

The Alarmed

The next group in the network of climate discourse are "the alarmed," the activist core of the climate movement.[19] In contrast to the techno-optimists who downplay climate dangers and trust in market forces, the alarmed freely express their fear of climate disaster and argue explicitly that achieving net-zero emissions requires unprecedented legislative and executive actions leading to "systems change." Systems change would include, for example, the transformation of our transportation system to enable electric-vehicle charging nationwide, high-speed electrified rail from coast to coast, and urban mobility centered on walking, cycling, and public transit. The alarmed believe that only governments have the regulatory and provisioning power, and the international standing, to catalyze and coordinate such transformations on a global scale.

At the same time, the alarmed observe, governments are currently captured by fossil-fuel producers. So, they argue, resolving the climate crisis will require not just technological or economic progress, but a political struggle—indeed, an epochal, even species-defining battle—between global civil society and the coal, oil, and gas interests who are themselves fighting to prevent the end of the fossil-energy era with every weapon they possess (long-standing incumbency, staggering wealth, well-developed propaganda platforms, and so on). To have any power in this battle, the alarmed maintain, civil society must raise a mass movement that will use democratic engagement and nonviolent direct action to force elected officials to break with fossil energy and commit, in an emergency mobilization, to decarbonizing the global economy as fast as equitably possible. And, earning their name "the alarmed," these commentators warn that if civil society loses this battle, and the required systems change does not occur in time, global heating will rise above a survivable limit, and climate change will, as it were, metastasize out of control, destroying the planetary conditions for ongoing human civilization.

One of the most prominent public figures among the alarmed is António Guterres, the United Nations Secretary-General. In July 2023—then the hottest month in recorded history, when the oceans for the first time became so warm that their temperature rose to 101.1°F, the temperature of a hot bath, around the Florida Keys—Guterres gave a speech announcing that "the era of global warming has ended; the era of global boiling has arrived."[20] Downplaying neither the violence of "global boiling" nor his feelings about it one whit, Guterres spoke about the human consequences of the carbon continuing to build up in our oceans and atmosphere: "children swept away by monsoon rains; families running from the flames; workers collapsing in scorching heat." Climate change "is here," he continued, "it is terrifying," and "it is just the beginning."

Nor did Guterres hesitate to point fingers at the people working as hard as they can to increase rather than phase out fossil-energy production. Rather, he zeroed-in on fossil-fuel producers and the banks and governments that support them. Fossil-energy companies must present the world with "detailed transition plans across the entire value chain" and stop spreading propaganda, Guterres said—"No more greenwashing. No more deception." Financial institutions must end their fossil-fuel lending and underwriting, he added, calling on brokers to "stop oil and gas expansion, and funding . . . for new coal, oil, and gas." Global North governments, he insisted, must put "a price on carbon" and push "the multilateral development banks to overhaul their business models." The "level of fossil fuel profits and climate inaction," he concluded, "is unacceptable." Indeed, for Guterres, the only acceptable option would be an emergency mobilization to eliminate greenhouse gas pollution and halt global heating. Going even beyond the targets in the latest IPCC report, Guterres called for "developed countries to reach net-zero emissions as close as possible to 2040" and for "emerging economies" to do it "as close as possible to 2050."[21] In sum, the whole speech exemplified the discourse of the alarmed: intense, explicitly emotional, focused on how climate change makes people sick and kills them, willing to call out bad actors and market forces like the profit motive, and committed to phasing out fossil energy at the greatest equitable speed.

Climate Scientists

Unlike the speech of the alarmed, the public communications of climate scientists are patently neither intense nor emotional. On the contrary. No matter how they may feel about climate change, climate scientists tend not toward public alarm but rather the reverse—toward cautious estimates that err on the side of less rather than more alarming predictions—what historians of science have called "the side of least drama."[22] Their nearly universal horror of exaggeration unites scientists into one category, even though practically speaking climate scientists display a remarkable diversity of political opinion, with some fitting easily into the techno-optimist category and others more liable to believe that resolving the climate crisis will require a political struggle. But this political diversity is masked by scientists' need to perform the cultural norms that have traditionally constructed scientists as reliable narrators. Scientists must signal their authority in part by presenting the results of their research with objectivity and dispassion. To this cultural injunction to be unemotional, add scientists' math-oriented sensibility, which recoils from more literary forms of speech like metaphor and allegory, let alone hyperbole, in favor of the unambiguous and literal language in which scientific findings are generally described, and you have a recipe for a communications style prone to understatement.

But climate scientists tend to speak dispassionately for reasons beyond the norms of their culture or their habits of mind. They are also responding to, and protecting themselves from, the decades of attacks—the decades of bullying, essentially—by the fossil-energy interests who called them "alarmists" any time they opened their mouths. Sir Robert Watson, a former Chair of the IPCC, once acknowledged as much in a public lecture: "We have to be very careful," Watson said, "if we ever have a strong statement that's later proven to be wrong, we will lose all credibility as a science community." This need for extreme caution led Watson to recommend that scientists "should always be slightly on the side of being conservative, otherwise we are going to get ripped apart by the climate deniers if we even make the most simple mistake."[23] As one of Watson's colleagues, the climate scientist Chris Rapley, put it more succinctly in a 2022 interview: "we were already being accused of being alarmists, so if we had told those stories [of tragedies like entire California towns burning to the ground], it would have undermined the level of trust in us."[24] Michael Mann, a prolific scientist and one of the world's most prominent climate communicators—who has been on the denialist firing lines since at least 2009, when his emails were hacked and misrepresented by fossil-energy interests to foment a manufactured scandal the news media called "climategate"—has described the extremely thin tightrope that scientists must walk in public. "Good scientists aren't alarmists," Mann explains, even if their "message may be—and in fact is—alarming."[25] Of course, maintaining this level of nuance, especially in the press, can be a real high-wire act.

Most of the time climate scientists do not stumble, but sometimes they do overstep and attack the discourse of the alarmed. After António Guterres proclaimed that the "era of global boiling has arrived," the NASA climate scientist Chris Colose took to X, once known as Twitter, to plead, with exquisite literal-mindedness: "There's no global boiling. Please don't let that be a thing."[26] A couple of days later, after he was criticized for not allowing public speakers to use metaphors, Colose tweeted that he hadn't been trying to fact-check the UN Chief—"I know most people don't think the oceans will boil (and wouldn't until temperatures in excess of 600 K, nearly Venus)," he said—but still he insisted that terms like "global boiling" are "cringe and let bad faith people get an easy laugh."[27]

Colose's self-defense received supportive replies from other climate scientists, including Katharine Hayhoe, a professor at Texas Tech and chief scientist at the Nature Conservancy, who tweeted an emoji signifying "100%."[28] Hayhoe is, perhaps, the most famous climate communicator; her special expertise is using conversations based on common interests to entice evangelical Christians and other right-wing partisans to care about climate change.[29] Hayhoe may have agreed with Colose because her communications style is designed to sidestep the polarized electoral politics of the climate crisis in favor of narratives about the availability of technological solutions meant to inspire a more apolitical "hope." Or perhaps, like Colose himself, she may simply have wanted to defend climate science from denialist ridicule. In any event, the whole debate over whether "global boiling" was or was not an appropriate phrase took place without anyone discussing the political elements or even the decarbonization targets in Guterres' speech.

Yet climate scientists attack rhetorical exaggeration not only because they want to depoliticize climate discourse. Even Michael Mann, who speaks out extensively against fossil-fuel-interest malfeasance, pushes back against hyperbole. As he put it in his book *The New Climate War*, there is "a danger in overstating the [climate] threat in a way that presents the problem as unsolvable, feeding into a sense of doom, inevitability, and hopelessness."[30] Mann worries, as do many of his peers, that exaggerating the dangers of climate change will make voters experience so much fear and despair that they'll tune out climate entirely. Their worry arises partly out of their reading the results of past social science research—such as one influential 2009 study showing that images of natural disasters and starving polar bears made low-income young mothers tune out thoughts of climate change—as if those results illustrated a natural law that every person on earth would shut down if they started to fear the harms of global heating.[31] But scientists' fear of exaggeration also grows out of their need to defend themselves from allegations that they've missed or hidden the fact that global warming has already spun out of human control. This latter accusation gets lobbed by the "doomers," the last node in this network of climate discourse.

Doomers

Doomers say not only that it is impossible to halt global heating, but also that such heating is going to lead to the collapse of civilization within decades.[32] Doomers come in three different factions. The first faction is composed of people who actively misrepresent the science—whether from ignorance or from some other malady—so as to insist that global heating has already triggered tipping points in the climate system and the planet is destined to become a Mad Max hellscape in our lifetimes. (Tipping points are long-onset but sudden events that transform ecosystems from one state to another, such as a mass die-off that finally turns a forest into a savanna after years of heat stress, repeated fires, and pest infestations. A good analogue to a tipping point is a coffee mug that falls off a table and shatters after being slowly pushed to—and then over—the edge.) But the consensus among climate scientists is that climate change has not yet set off tipping points.[33] Even the scientists who are most alarmed about the effects of global heating—in particular those who collaborate with Tim Lenton, the founder of the Global Systems Institute at the University of Exeter in the United Kingdom, who researches tipping points and other catastrophic climate events—maintain that "the rate at which damage accumulates from tipping—and hence the risk posed—could still be under our control to some extent." As an example, Lenton and his colleagues point to ice structures in West Antarctica that seem already to have started disintegrating in ways that could lead to roughly three meters of sea level rise, which would drown roughly 75 meters of land and infrastructure, measured from the shoreline, at coasts with slopes near the estimated global median value of 0.04, in addition to the substantial sea-level rise already expected as of 2023.[34] But they also note that this disaster would unfold over "a timescale of centuries to millennia," hardly fast enough to contribute to imminent social collapse. Still, scientists like Lenton and his co-authors strongly recommend that their colleagues do more research and communication about the possibility that even warming between 1.5 and 2°C might trigger major tipping points in order to, as they put it, "define that we are in a climate emergency" and strengthen the "calls for urgent climate action."[35] Yet most other scientists purposely avoid discussions about tipping points and their uncertainties. Not only are they strongly reluctant to opine on topics outside their specific area of research—and, indeed, most climate scientists do not study tipping points—but they also hope not to promote any false and counterproductive fatalism in doomers or, for that matter, the general public.

Downplaying catastrophic risks may not be so helpful in dealing with the second faction in the doomer camp, however. This faction is composed of people who sincerely believe that, whatever scientists may or may not say, the world is not going to eliminate its emissions and halt global heating because the *politics* of climate change are impossible. Young people are, painfully, very often in this camp. A study printed in the medical journal *The Lancet*, surveying 10,000 young people in ten countries, found that over 45 percent of people aged sixteen to twenty-five have experienced thoughts and emotions about climate change that "negatively affected their daily life and functioning." Notably, their distress is correlated not with hearing scientific misinformation (or, as Hannah Ritchie might put it, being told they're going to die) but with seeing that the adults in power don't seem to care their futures are being ruined—in other words with, as the study put it, "perceived inadequate government response" and "associated feelings of betrayal."[36] Their feelings of betrayal center on the climate crisis, of course, but arguably they are also fostered by a pervasive political cynicism, a lack of faith in the democratic process, which affects many people who came of age under neoliberalism. Whole books could be written about this problem.[37] Suffice it to say here that some millennial political leaders are trying to push back against this generational despair.

Representative Alexandria Ocasio-Cortez, who often speaks for her millennial and Gen Z supporters, freely expresses grief and fear about climate change, but she also explicitly rejects cynicism in general and doomism in particular. For instance, on an Instagram Live where she answered her followers' questions for over an hour, Ocasio-Cortez acknowledged that "the generation before us decided to consume our planet into the climate crisis, and they're still in charge and they won't stop." But at the same time, she affirmed: "I don't ascribe to climate doomerism because it serves no purpose . . . It doesn't help us get to a better place . . . I just really believe that climate doomerism and cynicism in general just leads you down a very dark path. And I get it—like, I have felt that way in the past, politically too. But

I never felt better than when I decided to check in my cynicism at the door and give it up and exchange it for: 'what if I tried?' And I would like to at least go down trying rather than go down just passively accepting having no agency in the world."[38] This is the message that young doomers need to hear most. Climate communication that downplays the dangers of climate change for fear of inducing despair will, at best, fail to address the political source of young people's anxiety—and could make young people feel all the more gaslit, as though climate scientists themselves were yet another constituency refusing to take their fears seriously. This is why Mann, for one, combats both scientific misunderstandings and political despondency by emphasizing both the "urgency" of decarbonization and the "agency" democratic citizens do have to redistribute power and force change.

Cynicism shades into the third and final form of doomism, where it becomes a pure form of nihilism that serves to justify one's own bare refusal to engage with the climate crisis. This is the doomism of the privileged—the Global North professionals confident that they can live out their days largely insulated from climate damages and secretly (or not so secretly) glad they will be dead before the proverbial shit really hits the fan. These people—often white, often male—not only misrepresent the science to say that the apocalypse is nigh, but also take their contempt for democratic politics as a badge of moral courage and intellectual sophistication, attacking climate activists for their supposed credulity in believing that the future is worth fighting for.

Case in point: Roy Scranton, an English professor at the University of Notre Dame, who publishes commentary in the *New York Times* and writes books with titles like *Learning to Die in the Anthropocene* and *We're Doomed. Now What?* Scranton insists that only "the deluded and naïve could maintain that nonviolent protest politics is much more than ritualized wishful thinking," as if protest politics has never led to change with respect to global heating (it has, as we shall see). But Scranton has no time for historical struggle. He imagines that "the story we're living is one of failure, catastrophe, suffering, and tragedy:

an out-of-control car careening off a dark road." Unsurprisingly, he grounds this extreme pessimism in misrepresentations of the science: "even if humans stopped emitting CO_2 worldwide today," Scranton says, "we would still face levels of warming over the next several decades that will . . . have a good chance of initiating runaway climate change."[39] This, happily, is patently untrue. For now, and for a few more decades at least, once emissions fall to net zero the planet will stop warming within three to five years.[40]

Doomers present a wicked challenge to climate communicators. Scranton's form of doomism can be as damaging to morale and the truth as any accusation of alarmism coming from fossil-fuel interests. It must be dealt with as a form of climate disinformation: debunked and given no further airtime. Yet the despair and grief of young people, which may look like Scranton's doomism (or, for that matter, the doomism of ignorance), needs in contrast a great deal of compassionate acknowledgment and fierce allyship, which would require not minimizing young people's fears. How to do both things at once, or know what to do when? Luckily doomers are a very small percentage of the electorate—as we have seen, YouGov has found only 10 percent of Americans believe it's already too late. The Program on Climate Communication at Yale University has found that only 13 percent of Americans agree with the belief that it's too late, with only 2 percent in strong agreement.[41] In truth, most Americans aren't even *alarmed* about the climate crisis. According to the Yale researchers, only about 26 percent of Americans can be classified as alarmed. The rest are concerned about climate change but assume incorrectly that its dangers are distant and therefore relatively low priority; or they're not sure global warming is even happening; or they simply never think about it; or they actively dismiss it as a hoax.[42]

The biggest challenge for climate scientists in the public eye, and indeed every person who wants to talk about the climate crisis more effectively, is to fight against the *general* anti-democratic fatalism of our current political era while seeking to use climate education and communication to move people out of the category of the concerned and into that of the alarmed—and then to mobilize the alarmed into taking disruptive or at least civil action to force their elected representatives to treat climate change as a looming catastrophe that requires an immediate emergency response. For, as we shall now see, it was a recent upswelling of climate alarm around the world that led to the first major piece of federal climate legislation in the United States. This alarm was a powerful force in climate politics—that is, before the forces of anti-alarm came together, sometimes inadvertently, to create a new consensus that "we" have averted the worst-case scenario so "we" can now exhale and look toward a damaged, but no longer destroyed, future.

From the Worst-Case Scenario to 3°C by 2100—The "Less-Alarmist Direction"

Starting roughly in 2018, fossil-fuel interests really started to highlight the word "alarmist" in their climate messaging. The Italian journalists Stella Levantesi and Giulio Corsi have found that between January 2016 and March 2020 the use of the terms "realist" and "alarmist" on Twitter grew by 900 percent, with the largest yearly increase recorded between 2018 and 2019.[43] An analysis by the media watchdog Public Citizen revealed that by mid-2019 the propaganda that climate scientists and activists were "alarmists," "hysterical," "chicken littles," or members of a "doomsday climate cult" had become a top climate message on Fox News.[44] And during this period then-President Trump repeated the talking point eagerly, as in a speech at the 2020 World Economic Forum, where he dismissed the climate warnings of the "alarmists" who, he said, want to "control every aspect of our lives."[45]

This drumbeat of the word "alarmist" was meant, it seems clear, to drown out the urgent calls for action that had been inspired by the 2018 release of the IPCC's Special Report on 1.5°C of warming (SR1.5). This report had been commissioned by the Parties to the United Nations Convention on Climate Change when they signed the Paris Agreement in 2015; it detailed the surprisingly large differences in harms between that Agreement's two targets—warming of 1.5° and "well below" 2° Celsius. It also set out a timeline for action that shocked policymakers out of their cowardly post-Paris reluctance to move against fossil fuels. To maintain even a two-thirds chance of halting warming at 1.5°C, the IPCC said, the world would need to halve its emissions by 2030, then only twelve years away, and draw them completely down to net zero by 2050. This deadline to halt warming at 1.5°C seemed so alarming not only because it was so awfully soon, but also because the report made clear how much worse things would be if policymakers allowed global warming to progress even 0.5° further, to 2°C. For just one example: at 1.5°C of warming, about 14 percent of humanity will likely be exposed to life-threatening heat on a regular basis. At 2°C that number more than doubles, rising to 37 percent, or approximately three billion people.[46]

To see how the IPCC report changed the tune of many world leaders, look at how the content and tone of António Guterres' discourse transformed after its release. In a 2017 speech Guterres actually allied himself with fossil-energy producers, promising to engage "all major actors, such as the coal, oil and gas industries, to accelerate the necessary energy transition," justifying his comity by contending, as he put it, that "we cannot phase out fossil fuels overnight."[47] By 2023, as we have seen, Guterres not only demanded that the fossil-fuel industry stop its ongoing "deception," but also called for the end of "oil and gas expansion, and funding and licensing for new coal, oil, and gas" as well as a decarbonization deadline even stricter than the one laid out by the IPCC itself: "as close as possible to 2040" for wealthy Global North countries. Once an industry-friendly centrist, Guterres became one of the most outspoken and hard-core fighters in the camp of the alarmed.

The IPCC's report also raised the alarm in the news media. Indeed, the *New York Times* editorial board published a statement about the report literally entitled "Wake Up, World Leaders. The Alarm Is Deafening."[48] This urgent tone had already been normalized to an astonishing degree by one journalist: David Wallace-Wells, who in 2017 had published an article in *New York Magazine* entitled "The Uninhabitable Earth," where he had detailed all the horrors of a world warmed by more than 4°C, including the possibility that humanity might become extinct. The article became *New York Magazine*'s most-read piece to that date, and in 2019 Wallace-Wells expanded it into a bestseller which vividly covered the planetary disasters projected for even lower levels of warming, making outright fear the emotion that seemed to dominate the climate conversation. (The *New York Times* op-ed heralding the book's publication was titled "Time to Panic. The planet is getting warmer in catastrophic ways. And fear may be the only thing that saves us.") Wallace-Wells was called an alarmist, of course, but he cleverly conflated "alarmism" and "alarm," embracing the insult as a name for an appropriate, normal response to the scientific facts, and thereby at least partly neutralizing the implication that alarmists were untrustworthy or emotionally unstable.[49] "I was called an alarmist, and rightly so," Wallace-Wells said later, "like a growing number of people following the news, I was alarmed. I am still. How could I not be? How could you not be?"[50]

Wallace-Wells maintained his credibility also because he was in fact following the science. When he wrote *The Uninhabitable Earth*, the world did seem to be on track to warm somewhere above 4°C by 2100. (This is about as much in 150 years, a geological instant, as the planet had warmed over tens of thousands of years to come out of the last Ice Age, when a glacier a mile thick covered what is now Manhattan. If you can imagine the difference between Manhattan covered by a mountainous glacier and Manhattan today, you can see how 4°C *more* of warming would remake the face of the planet, rendering much of the Earth uninhabitable for humans.) In 2017 and 2018, global emissions were understood to be tracking a socioeconomic model of the world's future emissions called RCP8.5. This model was one of the "Representative Concentration Pathways" (RCPs) that gamed out how different emissions scenarios—hypothetical stories about how much carbon dioxide the economy might emit in coming decades—would lead to different concentrations of CO_2 in the atmosphere and then to different levels of warming by 2100. RCP8.5 modeled the so-called "worst-case scenario"; it represented an emissions trajectory that would trap 8.5 watts per meter squared of extra energy in our climate system, leading to somewhere between 3.1°C and 5.1°C of heating by 2100.[51] At the time RCP8.5 was also characterized as the "no-policy baseline," which meant that it modeled the world taking no action to reduce emissions after whatever policies were in place in 2010. Although governments had pledged in Paris to try to halt warming at 1.5°C, even as late as 2018 their pledges had yet to be turned into policies in any meaningful way. So RCP8.5 was also called the "business as usual" warming pathway, and the world did seem to be on the road to ending the human experiment on this planet. Wallace-Wells argued that being honest about that fact and its consequences would help galvanize people—especially people with relative privilege like himself—by shaking them out of their blithe complacency.[52]

Wallace-Wells' insight that communicating the more terrifying dangers of climate change had political value seemed to be borne out by the self-professed motivation of a new movement of youth activists, who, horrified by the IPCC's Special Report and refusing to accept an uninhabitable Earth as their fate, rose up and forced governments across the world to pay attention. Some of these activists came together under the banner of "Fridays for Future" first advanced by the Swedish activist Greta Thunberg. After reading the IPCC's report in 2018, when she was fifteen years old, Thunberg began every Friday to go on strike from school and sit in front of her country's Parliament to protest its climate inaction. Within a few months, Thunberg's protest caught the notice of journalists, and the stark eloquence and incandescent moral clarity she expressed in interviews soon catapulted her into worldwide stardom, connecting her with youth activists around the world and giving her a platform at the center of global power. When she addressed the World Economic Forum at Davos in 2019, Thunberg told the attendees, including some of the world's richest, most powerful people: "Adults keep saying, 'We owe it to the young people to give them hope.' But I don't want your hope. I don't want you to be hopeful. I want you to panic. I want you to feel the fear I feel every day. And then I want you to act . . . I want you to act as if our house is on fire. Because it is."[53] Thunberg's alarming message—that the science is, frankly, terrifying and the only sane response is to phase out fossil fuels as soon as possible—mobilized unprecedented numbers of people. One Friday in 2019, millions of people in over 150 countries marched in a global climate strike, the largest ever, calling for immediate World War II–scale action against the climate crisis, and the very next Friday another several million marched again.[54]

In the United States, a youth activist group called the Sunrise Movement staged a sit-in at House Speaker Nancy Pelosi's office, demanding that Congress pass a Green New Deal—a comprehensive vision for full decarbonization proposed by Ocasio-Cortez and Senator Edward Markey (who had been on the front lines of the climate fight since at least 2009, when he helped introduce the American Clean Energy and Security Act, later known as "Waxman-Markey," the emissions-trading bill thwarted by Senate Republicans). The Sunrise Movement, which initially gave Joe Biden an F for his record on climate change, pushed the climate crisis to the foreground of American electoral politics and forced the Democratic candidate for president to reconsider his past positions on the issue. A shift in the electorate surely contributed too. By the end of 2019, the number of Americans who said they were alarmed by global heating had risen to 31 percent, and when these numbers were combined with the percentage of respondents who say they were concerned, nearly six in ten Americans reported feelings ranging from worry to outright terror of climate change.[55] The news media also continued to evolve in response to this change in public consciousness: for the first time in history, each 2020 presidential debate as well as the vice-presidential debate featured questions about global warming and the energy transition. In one of those debates President Biden even called climate change an "existential threat to humanity."[56]

Biden's newfound alarm about the climate crisis shaped his agenda even after he won the election. His first act after taking office was to rejoin the Paris Agreement, from which former-President Trump had withdrawn, and he centered his legislative agenda on expanding the American development, manufacturing, and uptake of clean-energy technologies. As we saw in the Introduction, in August 2022 he signed the Inflation Reduction Act (IRA), the first major federal climate law, which passed by only the slimmest of Democratic majorities in the Senate. The IRA invests public money into the deployment of clean energy and provides tax credits for utility-scale and distributed solar, wind, and other renewables, as well as nuclear and carbon capture, electric cars and trucks, and home-decarbonization machines like heat pumps and induction stoves.[57] This headway in the United States was also seen abroad with the development of the European Green Deal, a net-zero pledge by India, and China's release of an elaborate climate policy architecture called the "1+N" initiative (which I will discuss later in the book). By late 2022, the United Nations Environment Programme announced that the legislative and executive actions already in place across the globe would lead to emissions concentrations that would no longer produce 4°C or above, but likely 2.8°C of heating by 2100.[58]

This new heating estimate represented real progress. Although it partly reflected an updated scientific understanding of the planet's sensitivity to carbon dioxide concentrations—that is, the way the climate system responds to greenhouse gases collecting in the atmosphere—it also signaled that nations were finally beginning to translate their verbal Paris Agreement commitments into actual policies. This progress can be imagined as the first, hard-won nudging of a vast tanker, representing the global economy, which now needs to keep turning, set a new course, and accelerate. But the lower warming estimates were soon represented in the news media as something much less preliminary—as evidence, rather, that the world had successfully averted the worst-case climate scenario and that there was no further need to be so very alarmed.

This new narrative made its way into the news media via the Breakthrough Institute. Its preface was written in 2017, when a doctoral student named Justin Ritchie (no relation to Hannah) published his dissertation research arguing that the socioeconomic scenario underlying RCP8.5—which had modeled an over 500 percent increase of coal use in the late twenty-first century—was implausible in light of new estimates of the economics of coal extraction.[59] Soon Ritchie teamed up with the climate scientist Zeke Hausfather, then the BTI's Director of Climate and Energy, to write a blog post entitled "A 3C World is Now 'Business as Usual.'" In this post, Ritchie and Hausfather estimated that under "the [climate] policies and commitments [then] in place," the world would likely heat "well short of the 4C to 5C warming in many 'business as usual' baseline scenarios." And Ritchie and Hausfather used their revised estimate to argue not for further efforts to pass more policy, but for the efficacy of market forces: "Future technological innovation could continue to drive down the costs of both clean energy and electric vehicles even in the absence of new climate policy," they wrote. "It may be possible under an optimistic business-as-usual case to have as little as 2.5C warming by the end of the century."[60] Pushing this argument even further, Hausfather collaborated with Glen Peters, the Research Director for the Climate Mitigation Group at the CICERO Center for International Climate Research, and published an opinion piece in the prestigious scientific journal *Nature*, where they contended, more pointedly, that any talk of a planet heated to 4°C would be actively "misleading," because "the world imagined by RCP8.5" had become "increasingly implausible with every passing year."[61]

On their part, Hausfather and Peters advanced their arguments from the position that the cultivation of optimism, and the avoidance of doomism, best foments climate action. "Overstating the likelihood of extreme climate impacts can make mitigation seem harder than it actually is," they wrote; "this could lead to defeatism, because the problem is perceived as being out of control and unsolvable." At the same time, they stated clearly that global heating of 3°C would be a "catastrophic outcome."[62] Yet their joint polemic still sat uncomfortably close to accusations that scientists and activists were alarmists. Justin Ritchie, for one, went on to collaborate with Roger Pielke Jr., a political scientist at the University of Colorado who is a frequent Republican witness in congressional climate-change hearings; together they argued that the superseding of RCP8.5 proves that climate scientists purposely "misused scenarios for more than a decade" and that this misuse demonstrates a "pervasive and consequential failure of scientific integrity."[63] And Hausfather and Peters' work was soon weaponized by right-wing commentators as proof that governments had no need to pass any further policies constraining the production of fossil fuels or supporting the deployment of renewable energy.

Back in 2017 the conservative *New York Times* columnist and self-described lukewarmer Ross Douthat had allowed that "the closer the real trend gets to the worst-case projections, the more my lukewarmism will look Pollyannish and require substantial reassessment."[64] But in a 2022 column scolding youth climate campaigners for engaging in civil disobedience on the grounds that "fossil fuels are pushing the world toward climate apocalypse," as he put it, Douthat repeated the BTI researchers' language to say such campaigns were unnecessary: "climate change's existential risks have dropped meaningfully in recent years," Douthat intoned, "with worst-case scenarios becoming much less likely than before."[65] Over at the *Wall Street Journal*, columnist and editorial board member Holman Jenkins Jr.—who has written scores of denialist op-eds for his paper—cited Hausfather and Peters directly. Arguing that there is no need to enact further green policies, Jenkins claimed that his position is based on "what the science—the plain, recognized, consensus science—says about climate change: it won't be catastrophic." And what established this new "consensus science," according to Jenkins? The "drumroll moment" of "Zeke Hausfather and Glen Peter's [*sic*] 2020 article in the journal *Nature* partly headlined: 'stop using the worst-case scenario for climate warming as the most likely outcome.'"[66] Of course, Hausfather and Peters had said that our current business-as-usual outcome *would* be catastrophic. But the caveats tend to fall away when the language of optimism is echoed by lukewarmers across the network of climate discourse.

What collects these echoes into a cross-partisan consensus is their amplification by climate journalists, who powerfully shape the way that most Democrats understand the climate crisis. In this case, the story that we've avoided the worst-case scenario and are no longer headed for catastrophe was most clearly told, surprisingly, by David Wallace-Wells. Wallace-Wells first covered Ritchie and Hausfather's 2019 blog post in an article for *New York Magazine*. In that article, he pointed to all the nuances complicating Ritchie and Hausfather's polemic: the challenges of projecting technological and economic trends eighty years into the future; the possibility that the new emissions estimates did not fully account for planetary feedbacks which lead to higher levels of heating; the fact that projecting a level of heating doesn't tell you much about the harms of that heating. Yet ultimately Wallace-Wells recommended that "anyone, including me, who has built their understanding on what level of warming is likely this century on that RCP8.5 scenario should probably revise that understanding in a less alarmist direction."[67] (Leaving no stone unturned, the *Wall Street Journal*'s Jenkins weaponized this moment in Wallace-Wells' article too, crowing that even the author of *The Uninhabitable Earth* "was moved to call on fellow activists to revise their advocacy 'in a less alarmist direction.'")[68] Indeed, what now seemed most likely to Wallace-Wells was that the world could avoid catastrophe not by redoubling the political efforts of the climate movement, but simply by continuing "the blithe stumbling-down-our-current-path-blindly pattern of the last few decades."[69]

Wallace-Wells' 2019 article represented his first public attempt to get his head around the fact that the apocalyptic future he had feared and written about just two years earlier might no longer come to pass. As he said later on a podcast: "it's kind of miraculous, actually, how much different things look now than looked when I was writing that in 2018—we're very much not out of the woods, but the world is starting to pay attention . . . and that's what I've been wrestling with and reckoning with . . . how to tell that story [that says] we're not going to avoid the suffering but we're not going to all die . . . and how to balance those two narratives at once is really, I think, quite complicated." Grappling with this balance, Wallace-Wells next came down on the side of telling stories that would foment optimism and fight doomism. He apparently felt that the policy progress, along with the revised estimates of climate sensitivity, that had reduced warming from up to 5°C to around 3°C revealed how much power people have to determine the climate future: "[scenario] modeling is built on assumptions about the way the world will evolve," he said, correctly, "and it can evolve in different ways, in part because of how we take control of the levers of power." What he now wanted to convey, Wallace-Wells explained, was that "humans can change [the] future both with what we do to reduce emissions and also, on the other side, [how we] allow ourselves to protect . . . a more prosperous future for ourselves in a world that is nevertheless beset by many [climate] impacts."[70]

Accordingly in his next major attempt to tell this new, more complicated climate story—a long-form article and companion animated feature for the *New York Times Magazine*—Wallace-Wells described a future that would be full of disruption and horrifically bad for the world's poor, but at the same time more prosperous with respect to gross domestic product and, for Americans, in the end not that different from the world they live in now. Together his articles suggested that climate change no longer threatens the affluent; rather, they implied, the effects of burning fossil fuels will simply play out as an epochal injustice on what Wallace-Wells called "the plane of history: contested, combative, combining suffering and flourishing—though not in equal measure for every group."

Indeed, when "Beyond Catastrophe," the long-form article, cited the existential dangers of warming, it located those dangers far away from *New York Times* readers—with "Island nations" who call global heating "'genocide,'" or with "African diplomats" who call it "'certain death.'" But the article was able to portray the future as an unequal, unjust, but largely safe-for-Americans world because it mostly described a planet heated not 3°C by 2100, the estimated outcome of our current policy trajectory, but 2°C, a bit over the goal of the Paris Agreement. The companion animated feature, "A New World: Envisioning Life after Climate Change," went even further: it described *only* a world heated by 2°C.[71] But halting global heating at 2°C is hardly guaranteed. As Wallace-Wells himself put it, halting global heating at 2°C would require "a near-total transformation of all the human systems that gave rise to warming: energy, transportation, agriculture, housing and industry and infrastructure" (or, as he put it more politically in a later interview, "if we really fight maybe we can get [heating] down closer to two").[72] Still, when his articles pictured the future, they largely portrayed a 2°C world as what his readers could likely expect. When Wallace-Wells did invoke a 3°C future, he quoted Matthew Huber, a climate scientist who studies the temperature limit for human survival, calling that future a "win": "Some of my colleagues are looking at three degrees and going, oh my God, this is the worst thing ever," Huber was reported as saying, "and then someone like me is saying, well, I used to think we were heading to five. So three looks like a win."[73]

It's impossible to determine how much any one journalist or one newspaper influences the political conversation.[74] What can be said is that many Americans learn everything they know about climate change from the news media.[75] And that both the center-right and the center-left press were saying that "we" had avoided the worst-case scenario and we could now feel less alarmed. If after the 2018 IPCC report a new consensus had emerged—in which, for the first time ever, most climate scientists, activists, journalists, and Democratic politicians were all talking about climate change as an emergency—by the end of 2022 that consensus had fractured, with only activists and the most alarmed scientists spreading the message that all people, even Americans, were still facing mortal danger. It is perhaps telling that in September 2021 a record 33 percent of Americans said they were alarmed about global heating—nearly double the number who had said the same five years before—but by December 2022, this number had dropped a full 7 points to 26 percent.[76] To be sure, those who said they were no longer alarmed may well have been exhausted by the ongoing Covid-19 pandemic and simply picking their battles. Or, conversely, they may have tuned out the climate crisis in their jubilance that their social lives and the economy had largely returned to normal despite ongoing infections and deaths. Or, maybe, the previously alarmed had absorbed the news via osmosis, without paying close attention to the details, that the Democrats had passed historic climate legislation, the world was on a more positive climate trajectory, and heating 3°C by 2100 looked, comparatively, like a win.

The Catastrophic Outcome of 3°C

But is 3°C really a win? What will this amount of heating mean for people in the United States? How alarmed, exactly, should Americans be?

Well, 3°C of global heating would likely mean that for around three months of the year, the entire East Coast all the way to Maine, the Midwest to the Great Plains, and half of California would be so hot that being outdoors would put you at risk of death. Southern California, the Southwest, the Deep South, and Florida would be this deadly hot for nearly half the year.[77] And that's just the background heat. The effects of this heat in the climate system—what are blandly known as the "impacts" of climate change—should also be considered. And what the extreme weather already emerging across the globe, as well as the latest research, suggests is something quite frightening: the way our climate system is responding even to low levels of warming indicates that a 3°C world might look very much like the 4°C or 5°C world it had seemed like we had avoided. This is not to say that scientists have failed to predict what is happening in some way—their projections of how much the planet will warm at rising levels of carbon dioxide in the atmosphere have been entirely accurate—it is only to say that the extremes caused by warming are emerging on the worse side of the range of possible outcomes. As Andrew Dessler, the climate scientist and Director of the Texas Center for Climate Studies, explained in a 2021 television interview: "The models do an excellent job on the global climate, but as far as predicting these extreme events, it's looking like things are actually going to be worse than our worst-case analysis." ("For decades people called scientists 'alarmists,'" Dessler added, noting ruefully that "in the future that's going to end up being a really sad commentary on inaction.")[78]

Citing an event that suggested climate change was "actually going to be worse than our worst-case analysis," Dessler pointed to that year's Pacific Northwest heat wave, when temperatures in Seattle and Portland soared to 108 and 116°F, respectively, and British Columbia's temperatures broke records for three days in a row, topping out at 121°F, more than 40°F hotter than normal.[79] Public health officials later confirmed that this unprecedented heat killed nearly one thousand people.[80] Thousands more ended up in emergency rooms.[81] Scientists had not expected to see such deadly temperatures so soon in a region once known for its chilly mist and flannel shirts. As one study of the event noted, "heat conditions in North America in summer 2021 exceeded previous heatwaves by margins many would have considered impossible under current climate conditions."[82] Another study, agreeing with this assessment, also found that in a world 3.8°C warmer, such a heat wave would be an oven-like 5°C hotter.[83]

Such unanticipated heat extremes have continued to appear on land and in the sea all across the globe. Speaking about the 2022 European heat wave, which killed nearly 63,000 people (on the most developed continent in the world), the University of Oxford climate scientist Friederike Otto said: "all the climate models are underestimating the change that we see."[84] In 2023, while more than a hundred million Americans, nearly a third of the entire country, sheltered from extreme heat, dozens of people in Phoenix, Arizona—where temperatures rose to over 110°F and stayed there for a full month—were hospitalized in Intensive Care Units with third-degree burns after tripping and falling to the pavement, which had been made as searing as a frying pan.[85] In August of 2023, Buenos Aires surpassed 86°F; elevations of 4,000 meters in the Andes rose past 95°F; and other regions of South America crested about 100°F—even though it was winter in the Southern Hemisphere.[86] Meanwhile, nearly half of the world's oceans were suffering from sudden spikes in temperature that Michael Sparrow, head of the World Meteorological Organization's world climate research division, said were "much higher than anything the models predicted."[87] Although ice used to form easily in the once-frigid darkness of the South Pole, temperatures that winter never got low enough for seawater to fully freeze. "Unprecedented is a word that gets bandied around a lot, but it doesn't really get to just how shocking this is," said Australian climate scientist Will Hobbs. Losing South Pole sea ice at a mere 1.2°C of heating, he continued, "is very much outside our understanding of this system."[88]

The heat in the atmosphere and oceans was only one symptom of the fever spreading across the world. Climate scientists and economists once thought that Canada would thrive as the planet heated up, but Canada is now plagued by heat- and drought-fueled wildfires from June to October. In 2023 continent-spanning blazes destroyed 45.7 million acres of forest, more than double the previous record, while sending toxic smoke down into the United States and turning skies as far south as New York City an eerie, alien orange.[89] A native case of malaria was recorded in Maryland, and three people died of flesh-eating bacteria, a disease of warm tropical seas, contracted in Long Island Sound, one of the bodies of water bordering the Hamptons, the playground of the 1 percent.[90] In August of that year Tropical Storm Hilary, the remnant of a Pacific Ocean hurricane, flooded cities in Mexico and toppled trees and caused mudslides in Southern California.[91] Of course, scientists once thought that such storms were so unlikely in this region as to be virtually impossible, with one climatologist telling NASA in 2012 that "even if global temperatures were to rise six degrees, a hurricane in California would rank very low on the list of things we'd need to worry about."[92]

Frankly, none of these events had been seen as likely at less than 1.5°C of warming. And yet. One former Chair of the IPCC, Bob Watson, told a reporter that he is "very concerned." Although "none of the observed changes so far (with [then] a 1.2°C temperature rise) are surprising," Watson noted, "they are more severe than we predicted twenty years ago, and more severe than the predictions of five years ago."[93] While saying much the same thing, the current IPCC Chair, Jim Skea, confessed to being caught off guard: "it has surprised everyone," Skea told CNBC, "just the speed at which things have happened."[94] And David King, former Chief Science Advisor to the UK government, repeated Skea's assessment in an interview at the BBC: "We predicted temperatures would rise, but we didn't foresee these sorts of extreme events we're getting so soon," King admitted. "It's appropriate to be scared," he added.[95]

Climate scientists are even beginning to warn that the planet may be closer to a "termination-level event" than had been previously believed. Despite its apocalyptic name, such an event does not end life on Earth, but only a particular climate regime—an ice age, for example, which becomes an interglacial period. (Around 131,000 years ago the United Kingdom was cold enough to sustain glaciers in the Cotswolds, but after the climate flip called "Termination II," it warmed so much that hippopotami were able to wollow in what is now Trafalgar Square.) Such termination events take thousands of years to play out, but many include abrupt and rapid changes that happen over decades, such as the rapid 10°C heating over Greenland during the flip that brought about the pre-industrial modern climate.[96]

If the world does not stop powering the economy with fossil fuels, one such rapid change could occur much sooner than anticipated in the Atlantic meridional overturning circulation (AMOC). The AMOC is a massive system of currents, moving around fifteen million cubic meters of water per second, which runs through the entire Atlantic, from the Southern Ocean north toward the Arctic.[97] This current brings heat from the tropics to the European continent and regulates global precipitation patterns. As warm ocean water travels north along the East Coast of the United States, it slowly evaporates, becoming saltier; when that salty water enters the once-frigid outposts of Greenland and Iceland, it also cools. Since salty, cool water is heavier than fresh, warm water, as the current nears Iceland it begins to sink under shallower layers of the ocean, releasing its remaining heat into the atmosphere. This atmospheric heat makes Europe a temperate zone; were it not for the AMOC, cities like London, Paris, and Berlin would have the wintery climate of southern Canada, with which they share a latitude. But as the planet has warmed, the temperature of the North Atlantic has risen, no longer cooling the AMOC as effectively, and melting Arctic and Greenland ice has poured billions of tons of fresh water into the ocean, reducing the salinity of the current. All this has made the current slower and less likely to sink. Atlantic Ocean temperature patterns, including a "cold blob" suggesting the current is holding close to the surface, indicate that the AMOC is already getting weaker. These patterns may be shifting due to what scientists call "natural variability" rather than climate change.[98] But they may also be shifting due to climate change.

Scientists have long known that the Atlantic meridional overturning circulation has a tipping point, even if they're still not sure exactly where it is.[99] Yet, here too, it seems increasingly like it's coming sooner than anticipated. In 2013 the IPCC said that it was "*very unlikely* that the AMOC will undergo an abrupt transition or collapse in the twenty-first century" even under RCP8.5.[100] In their 2021 report, the IPCC downgraded their confidence in this projection to "medium" ("there is medium confidence that there will not be an abrupt collapse before 2100").[101] And in three studies released after the research for the 2021 report was compiled, scientists have found that the current is already giving off warning signals that it's in danger of collapse in the coming decades.[102] One of these studies, a Danish statistical analysis, found that if greenhouse gas emissions continue down the path to 3°C warming by 2100, the AMOC could shut down any time from 2025 to 2095, with the most likely time of collapse around mid-century, when a child born in 2010 would be in his forties and, one might hope, otherwise at the height of his family life and career.

If such a collapse were to occur, it would, as the IPCC put it dryly, "very likely cause abrupt shifts in regional weather patterns and the water cycle."[103] Over the next decades Europe would cool substantially—England would end up looking like northern Canada, but stormier—and the heat that would no longer be transported to the North Atlantic would end up staying in the tropics, turning densely populated cities into saunas where people simply could not live.[104] Global food supply would be severely threatened because Europe would dry out and the tropical rain belt would move south, weakening the African and Asian monsoons.[105] Meanwhile, sea level rise would accelerate over the East Coast of the United States, potentially stranding many billions in real-estate value.[106] As one author of the Danish study put it: "The implications would be devastating in terms of our ability to carry on living the way we do now, and to continue having agriculture in different places. You would probably have to change everything."[107]

Needless to say, not everyone agrees with the conclusions of these studies. The Danish study in particular received a great deal of criticism from scientists whose past research on ocean currents was contested by its findings. One scientist, with typical hair-splitting, acknowledged that "their tipping point analysis is robust and suggests the system is approaching a transition" but still insisted that he remained "unconvinced that this would be a catastrophic collapse." By contrast, Stefan Rahmstorf, Professor of Physics of the Oceans at the University of Potsdam, pointed out that "when multiple approaches lead to similar conclusions, this must be taken very seriously"—especially when the risk is of an event so catastrophic that it should be ruled out with 99.9 percent certainty.[108]

Many questions about tipping points remain open because the consequences of powering the economy with fossil fuels has no good analogue in planetary history. It is true that three million years ago, during the Pliocene era, the planet was about 3°C hotter than it was in 1850, so roughly the temperature to which it will heat again within decades if the world does not phase out coal, oil, and gas.[109] And we do know a lot about this hotter world. The Earth's oceans were so much higher that the eastern coastline of what would become North America ended about one hundred miles west of where it is currently. East Africa was heavily forested. The Arctic, now largely a frozen wasteland, was populated by giant camels, who roamed through lush woods of pine and birch.[110] It may seem safe to assume that if our economy heats the planet 3°C by 2100, Earth's current configuration would break down and be rearranged into something much closer to that Pliocene system. But even that is not certain. For the *rate* at which fossil-fuel use is adding CO_2 to the sky outstrips even the Paleocene–Eocene Thermal Maximum, the event with the highest release rate over the past sixty-six million years.[111] Peter Brannan, a research affiliate at the Institute of Arctic and Alpine Research at the University of Colorado-Boulder and the author of *The Ends of the World*, has even suggested that the rate of fossil-fueled carbon emissions exceeds that which contributed to the "global catastrophes that wiped out the majority of life on Earth," including "the greatest mass extinction of all time: a volcanic, CO_2-driven global warming cataclysm that nearly ended complex life a quarter-billion years ago." As far as we know, Brannan says, "humans are currently injecting CO_2 into the atmosphere ten times faster than any of these events."[112] To shift from the language of science to the language of religion: Lord, we know not what we do.

How to Talk about the Threat of Climate Change—And the Fight to Phase Out Fossil Fuels

It is perfectly appropriate to be alarmed. Given everything scientists are saying, given everything already happening at warming even below 1.5°C, it is reasonable—indeed, it is *sensible*—to feel frightened.

Fear of what may happen if we do not force policy- and decision-makers to end their support for fossil energy is not a symptom of alarmism. On the contrary. It's a sign that you are willing to look at the danger head-on and not look away. It is a sign of *courage*. You should talk about it as such.

To look at climate change clearly and not look away is to accept the reality of an epochal struggle that surely none of us would have chosen to participate in. But this is what everyone is facing: the world must phase out fossil fuels and remake the global economy as a net-zero-emissions system fully integrated into the biosphere. There's no way around it. The planet is going to continue to heat up until greenhouse gases stop being added to the atmosphere. Indeed, in some sense these debates about how much heating the world may or may not see by 2100 are a bit like the medieval wars over how many angels could dance on the head of a pin. Even if emissions projections end up being bang on, and the planet's temperature is exactly 2.8°C over preindustrial at midnight 2099, the planet will keep heating up and the climate will keep breaking down if we haven't phased out fossil fuels and brought anthropogenic emissions down to net zero. As Friederike Otto, the climate scientist at the University of Oxford, warns, the damage the world is already suffering is "not the new normal." The new normal will be, she says "what it is once we do stop burning fossil fuels."[113]

To preserve our safety, the world must stop burning fossil fuels now. Not in our grandchildren's or even our children's lifetimes. Now. The ramping down of consumption and the phaseout of production must begin immediately and be completed in a few decades. As we have already seen, the economy must halve its emissions by 2040 if the world is to preserve even a 50/50 chance to halt global heating at 2°C. And as António Guterres suggests, policy-makers in the Global North should accelerate that timetable of decarbonization for their already-developed economies, since it is becoming clear that even low levels of warming could lead to devastating harms. (The 2°C target was agreed to by world governments who were using now-outdated projections of extremes to inform their political cost-benefit analyses.) As we saw in the Introduction, the IPCC issued a stark warning in its latest report: "Any further delay in concerted global action will miss a brief and rapidly closing window to secure a livable future."[114]

Yes, this is all very frightening. It is important to pause and take slow, deep breaths if prickling sensations of panic begin to creep over your skin. You should calm yourself—but then also steel yourself. Have courage. And do not dwell on the terrifying news about climate change. To do that is to focus on what you cannot control. Instead—especially when you feel the most anxious—try to turn your attention away from the physical, planetary symptoms of running the economy on fossil fuels and instead focus on the powerful people who are working hard to preserve or even expand the use of coal, oil, and gas in America and around the world. These are the people maintaining the systems that are destroying the human future. The goal in shifting your attention to these people is to convert your fear into anger—to cultivate in yourself the galvanizing sense of righteous indignation that history and social science have shown best inspires political action.[115]

Remember that the American right remains entirely committed to promoting fossil energy, and, as we have already seen, they use disinformation and propaganda to justify their deadly energy policy and create false beliefs in their political base. Yet as we will discover in the next chapters of this book, even center-left people in powerful positions in government, research, business, and the news media also support the ongoing use of fossil fuels—even as they promote the expansion of clean energy technologies—and in so doing they say things that normalize fossil-fuel propaganda to justify their ambivalent but still dangerous policy positions. Keeping the language and the actions of all these people in view will help stoke a healthy outrage over fossil-energy interests' depraved indifference to the destruction of the only world known to support life. Focusing on the people and the politics of climate change will energize you to fight, helping you to take on and sustain the kinds of collective actions that will help remove these fossil-energy interests from power and replace them with policy- and decision-makers who will lead the transformation of our systems over the next decade and beyond.

As you become more practiced in accepting your fear and cultivating your indignation, you can then use the skills you've developed to energize other people to take on the fight to phase out fossil fuels. Try to deliver a three-part message. Make it clear that climate change will become catastrophic for everyone if the world does not phase out fossil fuels. Steer the conversation to the powerful people who are doing everything they can to prevent the transformation of the systems driving our global economy. Show how (as we will see in the next chapter) transforming those systems will have wonderful immediate benefits—clearer air; cleaner water; healthier lives; more leisure; everyone better-off from saving thousands of dollars every year on electricity, gasoline, and heat; communities renewed by new industries and ecosystem restorations; a sense of purpose and meaning in this human adventure besides just shopping—all quite aside from the basic but most precious gift of having a continued human presence in this universe.

Your goal in speaking out about the climate crisis in this way is to help people who might be concerned about global heating become more alarmed—and then to mobilize the alarmed into engaging in sustained, disruptive climate action. Of course, you will not move everybody with this three-part message. Indeed you may repel people who are generally disengaged from the climate crisis—not to mention centrist optimists—because it will be too much for them to take in at once. But that's ok.

As the political scientists Erica Chenoweth and Maria Stephan have found, just 3.5 percent of a population engaging in civil disobedience can be enough to force a revolutionary change in government.[116] These acts of disobedience are not perfect analogues for the climate movement, to be sure; they generally emerged in authoritarian nations and sought clear one-off goals such as regime change or territorial independence, rather than the more teleological goal of halting global heating, which would be the outcome of many lesser, instrumental triumphs along the way. And yet Chenoweth and Stephan's research is provocative and intriguing. 3.5 percent of the United States is around eleven million people—quite a large number, but by no means greater than the number of people who already say they are among the alarmed. Imagine eleven million people organizing against business practices that support oil and gas production, campaigning for candidates committed to passing comprehensive climate policy, or engaging in non-violent civil disobedience at high-profile media events: the political landscape could shift faster than Antarctic ice is currently melting. Already in 2023, 37 percent of Americans said addressing climate change should be a top priority for the president and Congress.[117] Mobilize a fraction of those people and we win.

Remember: the word "alarm" comes from the Italian battle cry *all'arme!*, literally meaning "to arms!" or "to the weapons!" One of the most powerful weapons you have is your voice. End the climate silence that gives fossil-energy interests cover. Talk about the climate crisis as much as you can.

Yet as we have seen, *how* you talk about climate change matters. And as we shall see in the following chapters, how you talk about "climate solutions" also matters. Whether it's the economics of clean energy, the role of technological innovation, or the geopolitics of climate cooperation, climate solutions, like the problem itself, are being misrepresented in fossil-energy propaganda—a propaganda that, there too, weaves itself into a centrist consensus that justifies expanding the production and consumption of coal, oil, and methane gas. To learn how that propaganda works and how to fight it—read on.

2

Cost

You probably read in your high-school textbooks that . . . extinctions were the result of asteroids. In fact, all but the one that killed the dinosaurs were caused by climate change produced by greenhouse gas.

—David Wallace-Wells (*The Uninhabitable Earth*)

When aggregate economic benefits from avoided climate-change impacts are accounted for, mitigation is a welfare-enhancing strategy (*high confidence*).

—The United Nations Intergovernmental Panel on Climate Change (*Climate Change 2022: Mitigation of Climate Change*)

Dealing with this existential threat to the planet and increasing our economic growth and prosperity are one and the same. When I think of climate change . . . I think of jobs.

—President Joe Biden
("Remarks by President Biden Before Signing Executive Actions on Tackling Climate Change, Creating Jobs, and Restoring Scientific Integrity")

The word "cost" is one of the most powerful propaganda terms in climate politics. It has long been used to justify Republican opposition to climate action. In 2001, when George W. Bush's Press Secretary Ari Fleischer was asked why the United States was withdrawing from the Kyoto Protocol, the first multinational attempt to halt global warming, he said that the treaty was "not in the economic interests of the United States" due to its "huge costs."[1] When Donald Trump announced that the United States would leave the Paris Agreement, he claimed that "the Paris Climate Accord [*sic*] is simply the latest example of Washington entering into an agreement that disadvantages the United States to the exclusive benefit of other countries, leaving American workers, who I love, and taxpayers to absorb the cost."[2] When Joe Biden announced as a 2020 presidential candidate that the United States would re-enter the Paris Agreement if he won the election, Wyoming Senator John Barrasso warned that re-entry "won't solve climate change" but "will raise Americans' energy costs"—and Texas Senator Ted Cruz echoed this charge, arguing that climate policies "will burden American families, manufacturers, and businesses with higher energy costs."[3] This GOP messaging is only amplified by right-wing media. Fox Business host Larry Kudlow, for instance, once opined on his show that Biden was "going to set strict, strict limits on carbon emissions within ten years [and] basically going to blow up the fossil fuel sector." This, he continued, was "going to cost a lot of jobs, just as the business tax hikes are going to cost a lot of jobs, just as the capital gains tax hike is going to cost a lot of jobs."[4] Kudlow's galloping repetition of "cost a lot of jobs . . . cost a lot of jobs" keeps the word "cost" echoing in our ears even as the various policies he is attacking fade into silence.

Oil and gas interests use the word "cost" so often partly because the term has a remarkable implicatory power to make people feel poorer. Asked in focus-group polls to "describe the first thing that comes to mind when they think of the word 'cost,'" respondents overwhelmingly answer that they imagine "a price" or "an expense" or an "amount of money you need to give up." Or they associate the word with "a burden" or "a sacrifice" or "money I don't want to spend." One respondent said the word made him inwardly lament "I don't have enough money!"[5] Fossil-fuel propaganda triggers these associations, stimulating the fear that government climate policy will hurt the economy and make everybody worse off.

The truth is rather the opposite. Decarbonization won't hurt the economy, I will show—climate damages will. Relative to those damages, decarbonization is not a cost, but an economic benefit. Indeed, it's a *windfall* that will not only preserve the priceless asset of our inhabitable planet, but also, here and now, raise the vast majority of Americans' real incomes.

This message contests the claims of fossil-fuel interests, of course. But it also flies in the face of the standard neoclassical account of climate economics, which, like fossil-fuel messaging itself, represents the costs of decarbonization as far worse for the American economy than the costs of climate damages, at least in this century. But this standard account, I will reveal, lowballs its estimates of the costs of climate damages—in part because it excludes from its models important scientific research on climate impacts and in part because it fails to account for the non-zero chance that global heating will lead to catastrophe, up to and including the infinite cost of our own extinction. With this infinite cost impossible to rule out, on balance achieving a net zero economy will always be worth it. Even in the immediate term, economists increasingly argue, phasing out fossil fuels will make most Americans more prosperous, easing their money worries and affording them greater resources to devote to their children's futures.

Fossil-fuel interests and centrist economists say that the costs of climate change are small and the costs of climate action are large. Here I will justify a new message: the costs of climate change are potentially infinite, and climate action is not a cost at all, but a social and economic windfall that will put money directly into Americans' wallets.

William Nordhaus and Cost-Benefit Climate Models

Climate economics was founded, almost single-handedly, by a Yale economist named William Nordhaus. A slim, dapper figure with a narrow face and elegant silver hair, Nordhaus designed the first cost-benefit climate model in the early 1990s, naming it the "Dynamic Integrated Climate Economy" model or DICE. (Punning on climate risks, Nordhaus often uses gambling metaphors in his work, entitling one of his books *The Climate Casino*, for example.) DICE combines elements of climate science with twentieth-century theories about economic growth to help policymakers estimate what Nordhaus calls the "optimal" price trajectory for a carbon tax—a trajectory which in theory balances the estimated costs of reducing emissions now with the benefits of avoided climate damages in the future. Using his model, Nordhaus intends to prevent policymakers from spending too much money on climate for too little benefit. And from an economic efficiency perspective, his approach makes sense. Just as you wouldn't buy a full suite of fire-fighting gear to extinguish, say, a grease-fire on your stove, governments shouldn't make huge investments in clean energy, according to this logic, if decades or even a century's worth of climate impacts will only damage the economy by a few percentage points on the margins.

Nordhaus published his first paper on DICE long ahead of the curve—in 1992, the very same year that the United Nations established its Framework Convention on Climate Change, the treaty that set the terms for international climate governance.[6] Over the decades Nordhaus has updated DICE, adjusting its equations and revising some of its inputs, but his policy advice has nevertheless remained the same. Nordhaus recommends that policymakers impose a low and only gradually increasing price on carbon, delaying the full decarbonization of the economy so as to maximize its productive capacity. "We must first remember," he counsels, "that capital is productive. Societies have a vast array of productive investments from which to choose. One investment is to slow climate change. But others will also be valuable."[7] Nordhaus suggests that individuals, businesses, and governments should invest their resources in financial products with higher short-term yields than climate projects. This way the economy can sustain what he defines as the optimal balance between the costs of climate damages and the benefits of growth. "The highest-return investments today," he writes, "are primarily in tangible, technological, and human capital, [but] in the coming decades, damages are predicted to rise relative to output. As that occurs, it becomes efficient to shift investments toward more intensive emissions reductions."[8] To achieve net-zero emissions too early, in Nordhaus' view, would be to miss out on the gross domestic product (GDP) growth that would arise from capital flowing to its most "productive" uses.

It is hard to overstate how influential Nordhaus' account of climate economics has been. In an extraordinary academic career—during which he has, to date, authored or edited some twenty books and scores of papers—he essentially produced the discourse with which economists, policymakers, businesspeople, and even activists imagine the costs of the climate crisis. In 2018 he won the memorial Nobel Prize for his scholarship. "What makes Nordhaus' contributions all the more notable," according to his colleague and co-author, the economist Kenneth Gillingham, "is the deep influence they have had on policy—something that cannot be said for every Nobel laureate."[9] The US federal government uses DICE, along with two other models developed on its structure, to price the social cost of carbon (SCC), the discounted marginal damages of one extra ton of carbon dioxide emissions.[10] Regulations with more than $1 trillion of benefits have been written with the SCC in their economic analysis.[11]

Yet it is striking that, from his position at the center of federal policymaking, Nordhaus has attempted to torpedo every international climate agreement for supposedly imposing, in Ari Fleischer's words, "huge costs" on America. Around the time Fleischer attempted to justify President Bush's pulling the United States from the Kyoto Protocol, Nordhaus criticized Kyoto as "highly cost-ineffective, with the global temperature reduction achieved at a cost almost eight times the cost of a strategy which is cost-effective in terms of 'where' and 'when' efficiency" (which is to say the "efficiency" of delaying fossil-fuel phaseout).[12] Nordhaus criticized the 2009 Copenhagen Accord, too, even though it was not even a binding agreement, but merely a statement affirming the general goal of halting global heating at 2°

Celsius. Nordhaus took to print to argue that "the simple [temperature] target approach is unworkable because it ignores the costs of attaining the goals."[13] Nordhaus was slightly more reticent to condemn the Paris Agreement, but in his 2018 Nobel Prize speech he repeated his objections to temperature targets as policy frameworks, again on the basis of their putative costs: "However attractive a temperature target may be as an aspirational goal, the target approach is questionable because it ignores the costs of attaining the goals."[14] Over the decades since he first ran the DICE model, Nordhaus' advice has remained the same: eschew low temperature targets and delay decarbonizing the economy. Specifically, Nordhaus recommends that policymakers time the implementation and increase of carbon taxes so that the world heats up not 1.5 or 2° Celsius, but 3° Celsius around 2100. As he told the Nobel audience: "in the DICE model . . . the cost-benefit optimum rises to over 3°C in 2100—much higher than international policy targets."[15]

One problem with targeting a global heating of 3°C as the "cost-benefit optimum" is that this target seems to ignore what climate science shows will happen if the world heats up that much that quickly. As we saw in the previous chapter, climate scientists warn that 3°C of heating would be "catastrophic."[16] This is why many climate scientists, including Johan Rockström, the director of the Potsdam Institute for Climate Impact Research, maintain that Nordhaus' approach is simply not aligned with the results of their research; it is "an unequivocal finding in the natural sciences," Rockström says, "that a 3°C warming is a disastrous outcome for humanity."[17] Yet Nordhaus sets this aside, arguing that we should delay full decarbonization and continue to use fossil fuels in the service of economic growth.

Nordhaus' argument may seem like a form of climate denial. Indeed Rockström once told the *Agence France-Presse* reporter Marlowe Hood that he "hear[s] this line of argument when confronted with the executive leadership at Shell, BP, ExxonMobil, the car industry and energy utilities"—which is why he thinks that Nordhaus provides "ammunition not only to climate sceptics, but to major actors that feel more comfortable with the status quo" who can say, well, "'if the optimal temperature for the economy is 3C, well then we can continue burning fossil fuels over the next century without any significant problems.'"[18] But Nordhaus is no climate denier. In his 2018 Nobel speech, he acknowledged that global warming "menaces our planet and looms over our future like a Colossus."[19] Indeed, Nordhaus' work provides an object lesson in the way that people's intentions become at least partly separated from the political effects of their speech. Nordhaus could, and surely does, intend to help deliver the best outcome for humanity (or at least the United States), given the reality of global heating, while still opposing halting global heating at 1.5 or 2°C. For the fault lies not in his intentions, but his assumptions. Those assumptions are grounded in the ideologies of the fossil-fuel economy that caused climate change in the first place.

The first such assumption is that economic growth will continue at its past rate even if the planet heats up by 3°C. "The United States and other economies will continue to grow over the next century in a manner roughly similar to that of the last century," Nordhaus intones reassuringly.[20] At the same time, with slightly less confidence, he suggests that cultivating this growth must remain everyone's priority. When he argues that our duty to our progeny requires that we not "decide for or tie the hands of future generations," Nordhaus means not that we must leave our children a livable planet, but that we must protect and grow our wealth to the greatest possible degree, so that we can bequeath that wealth to our children, giving them all the economic utility and freedom enjoyed by the rich. "Each generation," he writes, "is in the position of one member of a relay team, handing off the baton of capital to the next generation, and hoping that future generations behave sensibly and avoid catastrophic choices by dropping or destroying the baton."[21] In this allegory of an endlessly looping relay race, the "catastrophic" choice is to fail to bestow capital on the future. To crash the market, or perhaps to transform the economy into another form of production, thereby "destroying" capital, is, in Nordhaus' view, not to "behave sensibly." One may ask, though, how sensible it is to assume the race will continue when the stadium housing it is on fire.

Yet the fire is ignored, or downplayed, while Nordhaus largely takes it on faith that global GDP will continue to rise in the future no matter what happens to the climate. The faith is supported by his second major assumption—that growth is only marginally affected by climate damages, which are represented in Nordhaus' model as one-time shocks to the economy ameliorated by growth itself. Since DICE and its derivative cost-benefit models assume that climate damages are a simple fraction of GDP, they suggest that damages amounting to 2 percent of output, for example, could simply be offset by a 2 percent increase in output itself. With this offsetting mechanism, Nordhaus' model can show that more GDP is good, no matter how it's gained, for even if higher fossil-fueled GDP implies higher damages, an increase in GDP will still make the world better off.[22]

The more growth you assume, in other words, the less climate damages matter. They become a smaller fraction of the overall economy. Further, the more growth you project without climate policy, the more costly climate policies seem relative to future damages. This is especially the case in Nordhaus' model because DICE represents "climate policy" as a carbon tax and a carbon tax as an economic contraction. In this way, DICE implies that climate policy is something that policymakers should avoid enacting as much as they possibly can. Of course, Nordhaus does see the need to decarbonize eventually—after all, the world will continue to heat up until our economy stops adding greenhouse gases to the atmosphere—but his goal is to time full decarbonization to maximize capital accumulation. You might say that his approach is to play chicken, as it were, with climate change, stepping off the path of increased greenhouse gases in the atmosphere only once economic growth is immediately threatened to be toppled by climate damages—and not a moment before.

This moment comes so late for Nordhaus—at the point where the planet has heated up 3°C—because he assumes the cost of climate damages will be low. This is his third major assumption, and he sustains it by omitting from DICE some of the damages that climate scientists have identified as "the impacts" of global heating. He omits climate damages to labor productivity or to buildings or other infrastructure.[23] He omits damages to manufacturing, transportation, non-coastal real estate, finance, insurance, retail or wholesale trade, communication, and government services.[24] He omits damages caused by climate-influenced wars, coups, or social upheavals.[25] And, finally, he omits damages caused by climate-system tipping points and unlikely but possible civilization-threatening events that could be spurred by societal responses to ecosystem collapse.[26] The Columbia University economist (and former Senior Economist at the White House Council of Economic Advisers) Noah Kaufman says that to call any cost projection a best estimate, without accounting for these omitted impacts, "is a bit like counting up all the stars you can see in the sky and claiming the result is a best estimate of total stars."[27]

When Kaufman uses the word "estimate," he speaks literally. The numbers Nordhaus uses to represent climate damages are not quantities verified empirically, but postulates having surprisingly weak evidentiary foundation. It is fiendishly difficult, of course, to aggregate global damages from meta-analyses of sectoral scientific studies and project those aggregates into the future; in the early 1990s, when Nordhaus was developing DICE, there were few enough such studies that robust aggregation was nearly impossible.[28] So Nordhaus derived his damage estimates by surveying eighteen of his colleagues—nine economists (including, incidentally, former Secretary of the Treasury and President of Harvard Larry Summers), five natural scientists and engineers, and four other social scientists—about how much, according to their expert opinions, climate damages would cost in different temperature scenarios. Nordhaus also added his own opinions into the survey. The natural scientists, the actual experts on the effects of global warming, gave estimates that were twenty to thirty times higher than those of the economists, who had much greater faith than the scientists in the salvific power of money. Yet despite the fact the scientists were outnumbered by the economists about three to one, Nordhaus gave equal weight to each respondent's answer. This move tilted the study's bias toward the economists, enabling Nordhaus to find that "for most respondents the best guess of the impact of a 3-degree warming . . . would be '*small potatoes*.'"[29] Nordhaus then apparently input these "small potatoes" into his first version of DICE, and continued to use the same potatoes for decades, even as it became increasingly clear that the harms of global heating were baked in at lower temperatures than had been anticipated.[30]

Hence in 2013 Nordhaus pronounced that "the economic impacts from climate change will be small relative to the likely overall changes in economic activity over the next half century to century," with estimated impacts falling "in the range of 1–5 percent of output for a 3°C of warming"—a loss of income that would represent "approximately one year's growth for most countries spread out over several decades."[31] By 2018 Nordhaus had narrowed that range, projecting in his Nobel speech that climate change would cost a mere "2 percent of output at 3°C of warming."[32] When he updated DICE in 2023, he projected even fewer damages than that: "a 1.62 percent GDP-equivalent loss at 3°C warming."[33]

Not all economists agree with these estimates. Simon Dietz of the London School of Economics notes that Nordhaus' assumptions about growth and damage "arguably lead to gross underestimation of the benefits of emissions reductions."[34] The MIT economist Robert Pindyck is more blunt: Nordhaus' models and their derivatives, Pindyck says, "have crucial flaws that make them close to useless as tools for policy analysis: certain inputs . . . are arbitrary, but they can have huge effects [and] the model's descriptions of the impact of climate change are completely ad hoc, with no theoretical or empirical foundation."[35] A recent meta-analysis lists fully eleven critiques of Nordhaus' model, including one that condemns Nordhaus for having "outdated scientific understanding."[36] Yet Nordhaus still has significant influence over climate economics. One study notes that five of thirteen papers in a meta-analysis of climate damages were written by Nordhaus, with an additional four by the other two main modelers in the literature. This study pointedly argues that estimating climate damages is highly problematic when inputs "come from only a very limited number of pioneering authors who have published several estimates each," since the small size of the cohort produces "an exaggerated sense of knowledge about the temperature–damage relationship and a greater weighting of estimates from established authors in the field."[37] This is why in a brief for the Intergovernmental Panel on Climate Change (IPCC), an international group of climate scientists and economists warned that "climate policy recommendations based on the current framework seriously underestimate the economic value of climate damages."[38]

We can see this simply by looking at the historical record. Nordhaus projects a loss of 1.62 percent of output per year at 3°C of warming, but between 2015 and 2018, for instance, at a little over 1°C of warming, the United States lost almost 2 percent of GDP in weather-related disasters.[39] Hurricane Harvey, which hit Texas and Louisiana in 2017, and whose destructive rainfall was intensified 15 percent by global warming, accounted for larger economic costs than cost-benefit models assumed for an entire year.[40] The point is not only that such costs will continue to rise as long as the economy continues emitting greenhouse gases. The point is that intensified hurricanes, bigger fires, more frequent pandemics,[41] and relentless heat waves won't happen in isolation. Their costs will combine and compound in ways that Nordhaus does not model. And then, of course, if the world continues to use fossil fuels, the world will experience other climate damages that even scientists have not yet identified. "It is nearly impossible to completely account for the myriad possible interactions between the various components of the climate system," Gernot Wagner, a Columbia University economist, points out in a paper he co-authored with the climate scientist Cristian Proistosescu. "The world is riding this complex system into a state for which there is no good analog."[42] Statisticians, psychologists, epidemiologists, and economists call behavior whose full dangers are unknown "high risk" behavior. This behavior, far from being sensible, opens up the possibility of a truly catastrophic cost that might overwhelm any possible benefit the action affords.

Martin Weitzman and the Economics of Catastrophe

One of the first economists to recognize that the fossil-fuel economy is a form of high-risk behavior was a man named Martin Weitzman. A stocky, handsome New Yorker, built like a boxer and born on the Lower East Side, Weitzman spoke with such a thick New York accent that one colleague joked, "you'd think he was more likely to order the execution of a mafia don than lecture on the subtleties of prices."[43] Yet, with his accent intact, Weitzman ascended to the upper echelons of the Ivy League, receiving graduate degrees from Stanford and MIT, and teaching at Yale and MIT, before being hired by Harvard, where he remained for thirty years—gardening, when he wasn't working, and summering on a marsh island he bought off Cape Ann in Massachusetts. Weitzman may have been drawn to the economics of climate change because he knew something about unexpected catastrophes. When he was just a baby his mother died of tuberculosis. He was then abandoned by his father, who returned from World War II too broken to care for him, and placed in an orphanage. Later Weitzman was adopted by two schoolteachers who renamed him (he had been born Meyer Levinger) and raised him in Levittown, Long Island.

Despite his early traumas, Weitzman was a brilliant thinker. Gernot Wagner—who had been Weitzman's star student at Harvard and later became his co-author—said that Weitzman "was five steps ahead of you all the time."[44] He established the economics of cap-and-trade pollution trading, among other new forms of regulated capitalism that shaped governmental and business practices. But his greatest contribution was to the economics of climate change.

Weitzman recognized that climate breakdown comes with a surprisingly robust probability of worldwide catastrophe (which he defined as an event producing "a harm so great and sudden as to seem discontinuous with the flow of events that preceded it").[45] This probability of catastrophe is distributed in what is called a "fat tail," a particularly alarming kind of bell curve. A standard, "thin-tail" curve rises in a graph from the lower left and then flows symmetrically back down to zero on the right, making roughly the shape of a bell. A fat-tail curve, by contrast, does not flow all the way down on the right. Rather, it flows down part way but then extends outward as the events plotted on the graph get more extreme, making a shape that looks like the side-view of a humpbacked creature with a long, fat tail. A fat-tailed probability distribution assigns a higher likelihood to catastrophic events in the tail than does a thin-tailed distribution.[46] Nordhaus uses a distribution with a thin tail, which he "parameterizes" or "cuts off" so he can give a precise estimate in his damage function. But Weitzman looked at the science and saw that, as he put it, "if only gradually ramped-up remedies are applied," there would be a fat-tailed or statistically significant chance that global heating could lead to catastrophe—including the biggest catastrophe of all, the extinction of the human species. (This danger is flagged by the IPCC, which has noted, dryly enough, that "severe climate change might even lead to a catastrophic collapse of the population and even to the extinction of human beings.")[47] In some of his later work, Weitzman tried to put a number on this chance, proposing that if atmospheric carbon dioxide concentrations reached 700 ppm—in about 140 years at current emissions rates—the chance of human extinction would rise to a bit over 10 percent.[48] Weitzman felt passionately that "a greater than 10 percent chance" of "the end of the human adventure on this planet as we now know it" is simply "too high."[49] It helps to understand his passion when you consider that the chance of dying from Russian roulette is one in six, or 16.7 percent—not all that much higher than 10 percent, really. How many people would willingly take that fatal gamble?

But the bet that Nordhaus is placing on the survival of our species is even riskier than those numbers suggest. Weitzman's great insight was that the tail risks of climate change can never be adequately quantified. They are radically indeterminate, because they are emerging outside the full range of human experience—indeed of planetary experience. As we saw in "Alarmist," never has such a staggering amount of CO_2 been released into the atmosphere so quickly. And, if emissions continue, nobody can know with 100 percent certainty what the human outcome of this geological event will be. Or as Weitzman put it, in more economic terms, "I can't know precisely what these tail probabilities are, of course, but no-one can—and *that* is the point here."[50] This indeterminacy itself, this possibility of extinction that cannot be ruled out, means that the cost of climate breakdown is potentially infinite.[51]

Given the risk of losing everything, Weitzman recommended that policymakers should not play chicken with the climate and delay decarbonization, but rather adopt a generalized precautionary principle of "avoiding carbon emissions in the first place."[52] From a risk-management perspective it is better economics to spend more money on climate policy up front, with the expectation that as emissions diminish, and the probabilities of catastrophic outcomes decrease, society can safely spend less.[53] For even if you think, as Nordhaus does, that the climate crisis presents us with an "investment decision," Weitzman shows that this decision is, as he puts it in the jargon of finance, "an inherently fat-tailed situation with potentially unlimited downside exposure."[54]

Weitzman knew that his critique of the standard cost-benefit approach put the "policy relevance" of Nordhaus' model "under a very dark cloud."[55] He even argued that such an approach "falls apart as soon as we introduce the term 'catastrophe' and describe it as infinitely costly."[56] In one remarkable passage, Weitzman went even further than that, implying that Nordhaus' work was not only inaccurate and subjective, but blatantly misleading. "Perhaps in the end," Weitzman wrote:

> the climate-change economist can help most by not presenting a cost-benefit estimate for what is inherently a fat-tailed situation with potentially unlimited downside exposure as if it is accurate and objective—and perhaps not even presenting the analysis as if it is an approximation to something that is accurate and objective—but instead by stressing somewhat more openly the fact that such an estimate might conceivably be arbitrarily inaccurate depending upon what is subjectively assumed about the high-temperature damages function [and] about the fatness of tails and/or where they have been cut off. [And] explaining better to policymakers that the artificial crispness conveyed by conventional IAM-based CBA [cost/benefit analyses] here is especially and unusually misleading. [This] might go a long way toward elevating the level of public discourse concerning what to do about global warming.[57]

Nordhaus, on his part, responded to Weitzman's critique with a highly technical paper on the mathematical guts of his DICE model, in which he pointedly misprized the import of Weitzman's analysis. "There are indeed deep uncertainties about virtually every aspect of the natural and social sciences of climate change," he acknowledged, "but these uncertainties can only be resolved by continued careful analysis of data and theories."[58] Weitzman's point is that such resolution is not possible—at least not before it would be too late. Of course, it is hard to imagine that Nordhaus didn't *understand* the substance of Weitzman's work. It seems rather more likely that he just refused to take his colleague's point. In his 2018 Nobel speech, Nordhaus said flat out that "there is at this point no serious evidence of the presence of fat tails for the damage distribution," a tendentious claim that dismisses Weitzman's work entirely. But it's not like Nordhaus himself had never considered the possibility of catastrophe. Eleven years earlier, writing about how hard it is to quantify the desires of future generations, he had proposed that "perhaps we need to consider a model with uncertainty about preferences along with uncertainty about extinction."

It's entirely possible to imagine a future, he said, "where people come to love the altered landscape of the warmer world"—as if the chance that people might love a warmer world balanced out the risk of extinction.[59] But also, what about that warmer world? It is worth repeating here what I noted in "Alarmist": 3°C of global heating, Nordhaus' optimum, would likely mean that for around three months of the year, the entire East Coast of the United States all the way to Maine, the Midwest to the Great Plains, and half of California would be so hot that being outdoors would put you at risk of death. Southern California, the Southwest, the Deep South, and Florida would be this dangerously hot nearly half the year.[60] It's hard to imagine what kind love for warmth Nordhaus envisions people enjoying on such a planet.

Soon after Nordhaus won his Nobel Prize, Weitzman began to descend into a depression. In 2019 he ended his life by suicide, leaving a typewritten note saying he no longer felt mentally sharp enough to contribute to his field. Many of his colleagues reported later that he had been profoundly disappointed not to have received the Nobel along with Nordhaus. On his part, Nordhaus said that he had expected to share it with Weitzman. It's hard not to wonder whether Weitzman's disappointment might have been intensified by the implications of the prize committee's having ignored the importance of his work. Weitzman once told a reporter from the *New York Times* that he thought "we will keep drifting to higher and higher greenhouse gas concentrations until climate change is perceived as something catastrophic at a grass-roots level."[61] Maybe what his colleagues saw as the pain of a professional slight was at least partly the deeper agony of having failed to wake up the world.

One of the great tragedies of Weitzman's depression is that he apparently never saw how profoundly his work did influence climate politics. Without Weitzman, we would not have the 2° Celsius temperature target. As his Harvard colleague Robert Stavins said after his death, the Paris Agreement does not survive a cost-benefit analysis "unless you take account of Weitzman's fat tails."[62] And Weitzman's insights are now being integrated into the economics of climate policymaking, with the US National Academies of Sciences, Engineering, and Medicine calling for both "uncertainty characterization" and "scientific basis" (no more surveys of friends) in all future estimates of the social cost of carbon.[63]

How to Talk about the Cost of Climate Action

This "uncertainty characterization" should change not only estimates of the social cost of carbon, but also the way climate communicators think and talk about the economics of climate action. Now you should make it clear that the cost of *not* decarbonizing is potentially infinite, because scientists cannot rule out the frighteningly robust chance that high levels of heating could lead to human extinction. Any estimate of the costs of achieving and maintaining net-zero emissions must be weighed against this potential for unlimited downside exposure.

But this message alone will be only partly effective. Depending on your audience, invoking the risk of extinction might leave you open to an accusation of alarmism. And, in any case, this message requires listeners to project themselves into a potential future that might seem all too theoretical and distant. It's important to make the stakes of climate change entirely clear, but it's also crucial to be able to steer the conversation away from the question of extinction and toward people's more immediate concerns by providing some concrete climate-damage estimates that include what Nordhaus omitted. Here are a few.

Using already-observed responses to temperature variability—actual data—to revise and update Nordhaus' damage function (which they characterize, charitably, as a "rough estimate"), Marshall Burke of Stanford University and Solomon Hsiang of UC Berkeley have found that the costs of climate damages will be up to 100 times greater than prior estimates for 2°C warming. They forecast that unmitigated warming will reduce average global incomes roughly 23 percent by 2100 relative to scenarios without climate change.[64] Burke also examines Nordhaus' "optimal" warming targets in another important study, finding that 2.5 to 3°C global heating will lead to relative 15 to 25 percent reductions in per capita output in the same time-frame—roughly over the lifetime of a child born in 2020.[65] Swiss Re, one of the world's leading providers of insurance, has calculated even larger damages than that: 18 percent of GDP by 2050 in scenarios leading to a 3.2°C temperature increase.[66] And, in Asia, Chinese economists agree, showing in a 2020 study that 3°C of global heating will cost the world between $126.68 and $616.12 trillion over the next eighty years, compared to 1.5 or well below 2°C.[67] This onslaught of numbers can be dizzying, but here is a point that is easy to focus on: $616.12 trillion is nearly double the amount of money that currently exists in the world, and it can be expected to evaporate in 3°C of global heating.

So that is the first part of the message: the cost of climate damages has been significantly lowballed; economists are increasingly sure that they will be large and devastating; and, what is more, they warn that the infinite cost of our dying out cannot be ignored. The second part of the message is equally important: in the face of these costs, and even on its own merits, climate action is not a "cost" at all, but, for the vast majority, an economic windfall. As Burke notes, with some scientific reticence, decarbonizing quickly enough to halt warming at 1.5°C is likely "to result in substantial economic benefits" relative even just to 2°C of warming. He projects that halting global heating would be thirty times cheaper than letting the planet heat up to 3°C, with foregone damages in the tens of trillions of dollars.[68] Even economists using DICE have come to agree with this projection. Applying Burke and Hsiang's updated damage function and other scientific improvements to Nordhaus' model, two studies by independent teams of German economists—one from the Potsdam Institute for Climate Impact Research and the other centered in the Federal Ministry for

Economic Affairs and Energy—both show that the "optimal" temperature target is below 2°C.[69] When economists add what they call "natural capital" into DICE, they find that limiting temperature rise to 1.5°C is optimal.[70]

This flood of new research has, to be sure, weakened Nordhaus' hold over the economics profession, Nobel Prize or no. And this development has made fossil-fuel interests nervous. The power of Nordhaus' argument that ongoing fossil-fuel use is "optimal," and that temperature targets are unworkable due to their prohibitive "cost," has helped to justify political opposition to decarbonization. But the slow dismantling of Nordhaus' model, its ongoing reassembly as something grounded in reality, weakens some of the lobbying power that fossil-fuel interests have wielded over policymakers. (Even policymakers who believe in the reality of climate change are at risk of having their understanding of the crisis hopelessly muddled by economists who make decarbonization seem prohibitively expensive.) Thus even the American Enterprise Institute has started to excoriate "the climate left" for attacking Nordhaus, while insisting that "Nordhaus' absolute honesty and rigorous approach to economic analysis are beyond reproach." The "left" doesn't like Nordhaus, they say, because "DICE provides no support for the ideological attack on fossil fuels, on the freedom that they facilitate, or for such 'net-zero' policies as the Green New Deal."[71] And while it may well be true that the left doesn't like Nordhaus because he has opposed halting global heating at 1.5 or well under 2°C, it is not the "climate left" who has undermined Nordhaus' authority. It is economists themselves.

To repeat: a new generation of climate economists increasingly argue that relative to heating the planet over 2°C, rapidly phasing out fossil fuels and creating a net-zero economy by 2050 provides substantial economic benefits. That is a powerful message. But you may still be asked whether the world can actually *afford* to decarbonize so quickly and reap those benefits. The answer, by all accounts, is yes. Study after study shows that net-zero is affordable. A 2020 Princeton report shows that the United States could spend approximately what it currently spends on energy alone—about 4 to 6 percent of GDP—and decarbonize its entire economy by 2050.[72] A 2021 study by the Lawrence Berkeley National Laboratory finds an even lower price for the US decarbonization: between only 0.2 and 1.2 percent of GDP.[73] For perspective, consider that during the administration of Franklin D. Roosevelt, federal spending reached about 44 percent of GDP, leading to an extended period of upward mobility and prosperity for most white Americans. The US government will spend much less than that to create a net-zero economy and save the world. IPCC lead author Joeri Rogelj and multiple colleagues have calculated that, worldwide, just one-tenth of 2020's global Covid-19 stimulus, directed toward decarbonization each year for five years, would be sufficient to deliver the goals of the Paris Agreement and stop global warming well below 2°C.[74]

One reason that climate action is so affordable is that clean, safe energy has become cheap. In the past decade, the price of lithium-ion batteries has dropped almost 90 percent—despite the assumption in the dominant climate models that electric vehicles would remain more expensive than cars with internal combustion engines all the way to 2100.[75] The price of utility-scale battery storage, a partial solution to the "intermittency" problem of renewables, has also plummeted, falling 70 percent between 2015 and 2018 alone.[76] Onshore wind power is 40 percent cheaper than it was a decade ago.[77] And solar is now the cheapest source of electricity in history. Indeed, safe energy is cheaper overall than dangerous fossil-fuel energy almost everywhere on the planet.[78] New solar and wind farms save consumers nearly 40 percent compared to new gas plants and a whopping 66 percent compared to new coal plants.[79] (This is one reason why building out solar and wind in the developing world will best support increasing people's real incomes and alleviating poverty.) Taking a global view, the International Energy Agency has found that cumulative savings on fossil-fuel spending would total $12 trillion by 2050.[80] And since the price of fossil fuels doesn't follow learning curves, we should expect that fossil fuels will become overall only more expensive relative to safe energy in the future.

Many Americans do not know these facts or have been misinformed by fossil-fuel propaganda. Pew Research has found that 42 percent of Americans believe, incorrectly, that a systemic transformation to renewable energy would raise the price of household heating and cooling. Only 37 percent of Americans understand that it would actually lower household energy prices.[81] And by how much? A study from the Goldman School of Public Policy at UC Berkeley found that if America decarbonized its grid 90 percent by 2035, electricity bills would drop by 10 percent.[82] And this may well be an underestimate, since under Obama's Clean Power Plan, electricity bills shrank 14 percent between 2010 and 2020.[83] With his group Rewiring America, the MacArthur Fellow Saul Griffith modeled a national transition to solar-powered, fully electrified homes—in which induction stoves and heat pumps replaced methane-gas stoves and boilers and electric vehicles replaced gasoline-powered cars. He found that a fully electrified home and car could save the average household up to a solid $1,900 a year.[84] That is a lot of money. Vice President Kamala Harris often noted during the 2020 election that almost half of American families don't have even an extra $400 on hand for emergencies. Producing electricity with solar and wind power will improve the financial situation of these Americans especially. Imagine not having an extra $400 and then being able to save nearly five times that much annually. Such a rise in real income could be life-changing.

And thanks to phasing out fossil fuels and the machines that burn them, Americans will pay less not just for electricity, but also for health care. As the lead author of the Lawrence Berkeley study says, achieving net zero emissions "completely swings from net cost to net benefit when you think about damages from . . . air pollution and so forth."[85] Drew Shindell, Nicholas Professor of Earth Science at Duke University (and lead IPCC author), recently spoke to Congress about his team's latest research, testifying that, even leaving climate damages out of the models, decarbonizing quickly enough to halt global warming at 2°C would, in the next fifty years, prevent roughly 4.5 million premature deaths, about 3.5 million hospitalizations and emergency room visits, and approximately 300 million lost workdays in the United States alone (even though US air quality is relatively good).[86] Avoiding this sickness and death—not paying this cost of fossil fuels, you might say—amounts to a health-care and labor savings of over $700 billion per year, an amount of money that will pay for the majority of the transition to net-zero. And, Shindell emphasized, Americans will make this money from phasing out fossil fuels no matter what the rest of the world does. Running a version of their scenario in which the United States came into compliance with a 2°C pathway but the rest of the world continued to use coal, oil, and gas, Shindell's team found that unilateral American action would earn more than two-thirds of the health benefits of global action over the next fifteen years.

Of course, as affordable and lucrative as the transition to a zero-emissions economy really is, its up-front resource outlays, like those of any moneymaking endeavor, are real.[87] Yet we should call these outlays not "costs," but *investments*. They may not present much of an economic challenge—as we see in not only the new climate economics, but also the historical precedents of war and other crisis spending packages, including the global $12.2 trillion Covid-19 stimulus passed in 2020. Yet these investments do present a political challenge, even to governments that want to create net-zero economies. Nordhaus-style climate policy tries to drive decarbonization by levying carbon taxes, and these taxes really do increase voters' costs of living in the near term, sparking backlashes such as the French Yellow Vest protests of 2019. (Some carbon-price proponents advocate that the government ease the pain of higher prices by handing out rebates on a progressive scale, but rebates don't address the experience of consumer consumption being more expensive and painful in the face of inelastic demand for oil and gas, before the infrastructure for zero-emissions consumer choice is in place.) Increasingly, climate economists and policymakers are arguing against carbon taxes and proposing that governments enact robust industrial policy—a combination of standards, regulated credit policies, and direct grants into the safe-energy transformation, rebuilding infrastructure through public works projects and supplying cash rebates, zero-interest loans, or tax credits (on a sliding scale) to individuals and families so that they can decarbonize their households.[88] This was the approach that climate advocates and Democrats used to craft the 2022 Inflation Reduction Act (IRA), the first major federal climate legislation in the United States, which invests directly in clean-energy technologies and provides robust tax credits to consumers who want to electrify their homes. Of course, with a stated outlay of $380 billion over ten years, approximately 0.13 percent of US GDP, the IRA makes available only a small fraction of the liquidity that the net-zero transformation really needs, and it lacks standards that would give teeth to its provisions. Even so, it takes a new approach to investing in America's future.

Climate investments provide Americans with economic benefits. Comprehensive decarbonization policy would create millions of new jobs—jobs that would manufacture, deploy, and maintain solar and wind, necessarily domestic sources of energy. Not easily outsourced, these jobs could help revitalize towns whose main streets are shuttered and whose residents need to work two and sometimes three service jobs to pay their bills. The Goldman School/UC Berkeley study, for example, projects that climate action will create more than 500,000 jobs a year. As the Columbia University economist and Nobel laureate Joseph Stiglitz explains, "the war on climate change, if correctly waged, would benefit the country in the same way that World War II set the stage for America's golden economic era."[89] The unemployment rate of high-school graduates hovers around double that of people with a bachelor's degree, but even with a high-school diploma alone, Americans would be eligible to work in the new construction, conservation, care-giving, and adaptation trades that will be essential in a safe, zero-emissions economy.[90] The national project of phasing out fossil fuels and decarbonizing America could help repair the staggering income inequality and pervasive economic precarity of the neoliberal era.

Still, you are sure to be asked: where will we get the money to make these investments? Well, scholars give a variety of answers to this question. The first is offered by neo-Keynesian economists who espouse modern monetary theory (MMT), which describes the implications of the US federal government's power to print its own money. MMT economists argue that this power means, in essence, that the United States will always have funds for whatever spending Congress allocates, with the only limiting constraint being the danger of inflation.[91] Yet these economists also argue that money printed to pay for decarbonization would not lead to inflation, but would rather be absorbed by the purchasing of the new technologies and commodities (like induction stoves) required for a zero-emissions economy.[92] For progressive economists—Joseph Stiglitz and Princeton's Paul Krugman, for instance—who are skeptical about MMT in the context of international exchange, borrowing money seems smarter than printing it. As Krugman puts it: "We should be investing heavily in the transition to an environmentally sustainable economy . . . It would actually be irresponsible for the federal government not to engage in large-scale borrowing to invest in the future."[93] Be that as it may, many centrist economists prefer that Congress pay for its spending with tax revenue as it goes. And Congress can go a long way toward doing this by asking the astronomically rich to pay their fair share. US billionaires, a mere 650 people, could pay for over a year of the entire national transition to net zero with the extra money they made in the first year of the Covid-19 pandemic alone.[94]

Any of these methods would effectively distribute decarbonization investments so that the vast majority of individual consumers wouldn't really feel them (billionaires have so much money that they wouldn't really feel higher taxes either). This claim may seem implausible, but think about the financial crisis of 2007–8, for example. The Government Accountability Office reports that in that crisis the Federal Reserve spent $16 trillion to stabilize the financial system—which, according to the estimates cited earlier, is close to the amount of both public *and* private money that will be needed to decarbonize the US economy.[95] Can any of us say exactly how much of that $16 trillion came out of our wallets in particular? Probably not. And yet the bankers came out richer, because they got an infusion of public investment which they leveraged into equities and other financial products. Congress could infuse the rest of us with public investments, which we can leverage into sectoral decarbonizations that will help preserve an inhabitable planet and put money into our pockets, to boot. In the end, much climate funding may well come from the private sector, of course, but the federal government must create the conditions in which a net-zero economy will root, develop, and flourish.

Yet there's a hitch to this. Although most Americans will be much better off once the economy is decarbonized, fossil-fuel interests and their political allies stand to lose a great deal of money. To meet even a 2°C target, a third of oil reserves, almost half of methane gas reserves, and over 80 percent of current coal reserves must remain in the ground.[96] This unburnable carbon is currently valued as high as $3.3 trillion,[97] of which $350 billion is owned as equity in the United States—one-third by the bottom 90 percent and the remaining two-thirds split roughly equally between the next 9 percent of asset holders and the top 1 percent (who of course have many other assets).[98] To halt global heating at 1.5 or 2°C these fossil-fuel assets must be stranded—devalued, written down, or converted to liabilities.[99] And fossil-fuel companies and state producers, not to mention many wealthy people, are doing everything they can to prevent that from happening.[100]

Although pretending to embrace the Paris Agreement, and spreading propaganda about being poised to deliver carbon-removal technologies (they call these "innovations," as we shall see), fossil-fuel companies actually allocate only about 4 percent of their upstream capital investments to climate solutions—mostly to new bioenergy, a marginal climate solution at best.[101] Meanwhile, they lobby to block the enactment of robust climate policy, while spending hundreds of billions annually to discover and extract new fossil-fuel deposits. The United Nation's 2021 "Production Gap Report" found that fossil-fuel suppliers have planned to produce by 2030 more than double the amount than would be "consistent with" halting warming at 1.5°C and nearly 45 percent more than would be "consistent with" halting warming at 2°C.[102] In other words, fossil-fuel companies and petrostates are fully planning to heat the world up around 3°C by 2100. They will carry out their plans unless democratic governments stop them.

It will be very challenging to manage the decline of the fossil-fuel industry. (Indeed, climate advocates cannot promise decarbonization will be easy—just that it will be affordable.) In the United States, Congress could buy out investors, establish a national fossil-fuel company, and then legislate an incremental ban on private production while delivering economic justice to oil and gas workers. At the very least, within a market framework, it could stop artificially inflating the value of oil and gas majors with subsidies while legislating climate-risk disclosure, in order to encourage finance to reallocate capital to climate solutions. In either case, one fact makes the challenge easier. The United States does not rely on oil revenue to fund its public services (as do Iraq, Nigeria, Oman, and Saudi Arabia). And even as their profits can skyrocket after geopolitical shocks, such as Vladimir Putin's 2022 invasion of Ukraine, the fossil-fuel industry is no longer a solid cornerstone of the American economy. Yes, they can return record dividends to their investors, but at the same time they are shedding jobs. Only about 600,000 Americans now work in mining and oil and gas extraction, compared to the 1.8 million who work in real estate—a sector directly threatened by climate damages, especially on the coasts.[103] Fracking promised to deliver Appalachia from the economic devastation of the outsourcing of the American coal and steel industries, but it has done no such thing. The Ohio River Valley Institute has found that the twenty-two Ohio, Pennsylvania, and West Virginia counties responsible for 90 percent of the region's oil and gas production saw their share of the nation's jobs, personal income, and population all decline after fracking began in their communities.[104] And due to increased automation, the number of workers required per extraction project will soon fall by 20 to 30 percent, even without a transition to renewables.[105] The oil and gas worker is fast becoming a mythical figure in American culture, akin to the cowboys in 1950s film and television, representing a nostalgic invocation of a past expansion of American imperial power. Today the fastest-growing job in America is solar installer.[106] Approximately 472,000 people work in the solar and wind industries already, and the clean-energy revolution is just getting started.[107]

Miners and oil and gas workers know the future lies elsewhere. In April 2021, in a historic event, the United Mine Workers of America, America's largest coal union, announced that it would accept a transition away from fossil fuels in exchange for new jobs in renewable energy. Its proposed transition plan asks the federal government to encourage renewable-energy development in former coal country, retrain miners to work in safe energy, and guarantee their wages, health insurance, and pensions during the transition.[108] This is exactly the kind of systems-change approach to climate policy that would be enabled by a nationalized, managed phase-out of fossil fuels, under principles set out by the Green New Deal.

A systems-change approach to phasing out fossil fuels will not only create new industries and new jobs, but also ensure that no worker gets left behind in the transformation. It will ensure that America decarbonizes the economy at the scale and the speed required to halt global heating to well below 2°C. And it will guide Congress into making the investments into climate action that will raise most Americans' real income immediately, leaving them healthier and more prosperous than before. Fossil-fuel interests and supporters of the status quo want to prevent the federal government from making these investments and coordinating transformative systems change. That is why, insofar as they appear to support climate policy, they do so, as we shall see in the next chapters of this book, to extend their social license to operate—or, sometimes, to impose a moratorium on climate regulations in exchange for a nominal carbon tax which would impose costs directly on consumers and weaken their support for climate action. Indeed, weakening support for climate action is the very goal of the propaganda emphasizing the putative "cost" of climate policies at every turn.

To help neutralize this propaganda, you can emphasize instead the costs of climate *damages*, using alarming examples from the extreme weather that is already emerging all around us. And you can connect these examples to the broader economic point that if the world does not phase out fossil fuels, the cost of climate change will potentially be infinite, inasmuch as unchecked global heating threatens to end the human adventure on this planet. Then, you can stand on ever-growing heaps of economic evidence to say, happily, that decarbonization is a *windfall* that will put money directly into most Americans' pockets.

Talking about the economic dimensions of the climate crisis in this way is not only accurate, but also politically effective. Calling decarbonization a windfall for the majority both exposes and counters the propaganda that Americans' prosperity relies on fossil fuels. Your message will convey that in reality fossil fuels are poised to destroy the progress they once helped create, and that the vast majority of us will be better off—healthier, and wealthier—without them. Finally, helping to build a new worldview for the post-carbon era, this message will introduce a new understanding into the language of climate politics: the understanding that an economics which accounts for ecological considerations can actually increase human flourishing.

3

Growth

> There are good Darwinian reasons why cultures without any future-orientation should fail to survive very long in the course of history.
>
> —Robert Solow ("Is the End of the World at Hand?")

> Land provides the principal basis for human livelihoods and well-being.
>
> —The Intergovernmental Panel on Climate Change (*Special Report on Climate Change and Land*)

> The growth of the modern economy might turn out to be a colossal fraud.
>
> —Yuval Noah Harari (*Sapiens*)

The word "growth" is a core term in the language of climate politics. As we just saw, climate economists estimate the costs of climate damages relative to the amount of economic growth they project into the twenty-first century, and by and large they assume that such growth will continue no matter how hot and chaotic the planet gets. Their belief in a future of continued economic growth enables another idea sometimes used to justify sustaining the fossil-fuel economy: the idea that economic growth is itself a climate-change solution, a form of environmental protection that will shield the prosperous from climate devastation. This belief is so bipartisan, so ubiquitous, that it's not quite accurate to call it propaganda. It's best understood as a *myth*. This myth is often taken for reality by both fossil-fuel partisans on the right and climate-conscious progressives on the center-left.

So, for example, the right-wing think tank The Heritage Foundation, which has spread fossil-fuel propaganda since at least the 1990s, argues that "economic growth" is "one of the most important factors for maintaining a cleaner environment," since as a country's economy grows, "the financial ability of its citizens to take care of the environment grows, too."[1] In a similar vein, the conservative *New York Times* columnist Bret Stephens, whose past climate denial engendered a storm of protest when he was hired by the paper, wrote a sort of *mea culpa* in 2022, when he finally acknowledged that global heating is real—but also proclaimed that "economic growth should be seen as an ally in the fight against climate change, because it creates both the wealth that can mitigate the effects of climate change and the technological innovation needed to address its causes."[2] Ezra Klein, Stephens' progressive colleague at the *Times*, likewise sees growth as an ally in the fight against climate change, both because economies of scale help make green technology cheap and because, to his mind, the "political appeal" of the net-zero emissions world relies on a politics of abundance, rather than on what he calls "sacrifice." And, while he recognizes the vast injustice of global inequality, and the threat climate change poses to the poor, Klein maintains, as does Stephens, that "richer people and countries will buy their way out of the worst consequences [of global heating], often using wealth accumulated by burning fossil fuels."[3] In a similar vein, while excoriating the climate-change film *Don't Look Up* for its supposed alarmism, Eric Levitz, a senior correspondent at *Vox*, avers that "it is possible to imagine the top 10 percent of America's income distribution living *relatively* comfortably in a two- (or perhaps even three-) degree-warmer world."[4] And even Adam McKay, the avowedly democratic-socialist director of *Don't Look Up* itself, enacts the belief that wealth will save the rich from climate apocalypse. McKay ends his film with Earth on fire and climate-denying billionaires launching themselves into space to colonize a new planet—where, with delicious poetic justice, they all promptly get eaten by giant extraterrestrial birds.

What justifies the belief that wealth will enable the rich to escape climate apocalypse? An important part of the answer is the economic modeling of the relationship between gross domestic product (GDP) growth and global heating. As Ezra Klein has pointed out, "no mainstream climate models suggest a return to a world as bad as the one we had in 1950"—a year when white Americans, at least, were largely doing well.[5] But what enables "mainstream climate models" to project continued economic growth even into the hottest futures? It is important to clarify at the outset that growth projections do not arise from a systemic assessment of all the interactions between the economy and the biosphere through which it emerges. Such an assessment would be impossible. What climate economists do is study the effect of one variable, such as heat, on a growth trajectory that they take as a given—that is, "exogenous" to the model, to use the term of art. And that growth trajectory is explicitly decoupled from ecological considerations. Yet at the same time, the modeled decoupling of the economy and the biosphere is taken not merely as a heuristic, but as a true story about how the world actually works.

I will reveal that there is no real basis to this story, which is implicitly activated any time commentators talk about the relationship between climate change and growth. To take the true measure of the ways that climate change could harm economic growth, economists will need to better integrate ecological considerations into their theories. Without this integration, it will be impossible to see clearly whether wealth might continue to be protective as the planet leaves the climate niche in which complex economies have developed over the past 10,000 years.

To be sure, this message of doubt belies the confidence of many climate economists, and the apparent evidence all around us, that wealth does protect nations and people from climate change. But projected onto the future, this confidence, I will show, rests on two unfounded dogmas: first, that technology can always substitute for nature and, second, that wealthy societies have nearly limitless capacity to adapt to global heating. These are the dogmas that sustain the myth of economic growth itself as a form of salvation from climate change. It's long past time to understand the flimsy origins of this myth and start circulating a new climate message grounded in reality: it is quite likely that ruining the planet will also ruin the economy.

Robert Solow's Theory of Perpetual Growth

If the economics of climate change was founded largely by William Nordhaus, the neoclassical economics of growth enabling Nordhaus to project rising GDP even in a world heated 3°C or more was created by Nordhaus' PhD supervisor at MIT: a man named Robert Solow, himself a Nobel laureate and towering figure in American post-war economics. Over the course of a five-decade career, Solow indelibly shaped academic and popular ideas about the relationship between economic production and the natural world. His foundational paper, "A Contribution to the Theory of Economic Growth," published in 1956, still undergirds the mainstream understanding of economic development. A Brooklyn boy—like Martin Weitzman—Solow went to Harvard on a scholarship when he was only sixteen, leaving two years later to volunteer for World War II, and returning to Harvard and a stint at Columbia before joining the MIT faculty in 1949. Tall and laconic, with a wry, self-deprecating sense of humor which enlivens even his scholarly writing, Solow taught not only Nordhaus, but also three other Nobel laureates, including Joseph Stiglitz, and was himself awarded the Nobel as well as the John Bates Clark Medal, the National Medal of Science, and, by Barack Obama, the Presidential Medal of Freedom.

Solow helped overturn more than a century of economic orthodoxy, rewriting the so-called "classical" theories of growth advanced by eighteenth- and early nineteenth-century economists like Adam

Smith and David Ricardo. These enlightenment thinkers saw the building blocks of the economy as three "factors of production": labor, capital, and land. And they understood land to be scarce, which is to say available only in finite quantities. Because one of their factors of production was scarce, classical growth theorists imagined economic growth would eventually come to an end. At some point farmers would not be able to grow enough food to feed an endlessly growing population, and at that point the labor force would stagnate, and the economy would resolve itself into a steady state where it would kind of bump around the same level of production indefinitely and revenue would, at best, cover only the costs of capital depreciation.

By the time Solow constructed his neoclassical theory of growth in the 1950s, twentieth-century economists such as Keynes and Schumpeter had already explored the ways that fiscal policy and innovation could enable wealth-creation by establishing new markets and investments quite outside of the interactions between labor and land. But it was Solow who managed to decouple land from other factors of production and theorize perpetual economic growth without biophysical limits. He did this by building a mathematical model that replaced land with a new production function. (A production function is a mathematical construct that performs operations on input variables, like labor and capital, to generate output values.) He gave his function the optimistic name "technological possibilities," and he stipulated that it would have a constant positive rate of return. This constant positive rate of return would multiply the outputs of labor and capital indefinitely. In short, Solow kept part of classical growth theory—labor and capital interact to produce total output, some consumed, but some reinvested, leading to capital accumulation—but he also introduced a mathematical operation that enabled him to model economic growth freed from the limits that land scarcity places on agricultural production.

Solow wrote out the mathematics of his model in "A Contribution to the Theory of Economic Growth." His paper became extremely influential because its theory seemed to describe post-war American economic growth with remarkable accuracy. Its descriptive power is perhaps not entirely surprising in light of twentieth-century technological developments—the transition from horsepower to cars to planes to space flight, the rapid industrialization of the American heartland, and especially the dissemination of fossil-gas–based fertilizer that exponentially increased agricultural productivity, seemingly freeing humanity from the Malthusian trap. But the problem with its success is that Solow's model was taken not as historical commentary, but as scientific law. When Simon Kuznets, for example, accepted *his* Nobel prize for, among other accomplishments, standardizing the concept of gross national product, a precursor to GDP, he defined growth itself in Solow's terms—as "growing capacity based on advancing technology and the institutional and ideological adjustments that it demands."[6]

It's important to emphasize here that despite its power to describe twentieth-century economics, there was nothing empirical about Solow's new production function—no scientific fact in the world established that something as ephemeral and forward-looking as "technological possibilities" could produce constant rates of return indefinitely. But of course in his early work Solow was not trying to discover scientific facts; he was hypothesizing about how perpetual growth might actually work, while calling his hypothetical scenario a contribution to a "theory" of growth. And, as he wrote, "constant returns to scale seems the natural assumption to make in a theory of growth."[7] (To his credit he also wrote, in the very first sentences of his paper no less, that "all theory depends on assumptions which are not quite true. That is what makes it theory.")[8]

Yet one not-quite-true assumption begat another. The assumption that technological possibilities will always provide constant returns to scale led directly to the assumption that agricultural production has no influence on growth rates. As Solow put it, using the language of calculus: "the production function is homogeneous of first degree. This amounts to assuming that there is no scarce nonaugmentable resource like land." Indeed, he suggested, a proper theory of growth *must* assume there is no scarcity of land, or even no planetary boundary at all: "the scarce-land case," Solow pointed out, "would lead to decreasing returns to scale in capital and labor and the model would become more Ricardian" (which is to say, subject to limits).[9] In other words, Solow jettisoned planetary boundaries to model a theory of perpetual growth. And insofar as he wondered what setting aside planetary boundaries in his model might say about the real economy or the world at large, he confined his speculations to a footnote, where he acknowledged that endless growth actually *does* depend on the planet—specifically on the endless production of new arable land via extraction from forests. "One can imagine," he noted, "the theory as applying as long as arable land can be hacked out of the wilderness at essentially constant cost."[10] In this footnote, where Solow imagined his theory as actually "applying," land re-enters the picture as a natural resource whose amounts do in fact need to grow to enable the growth supercharged by technological possibilities—that airy, dematerialized optimism driving reinvestment and capital accumulation. Here growth is still dependent on arable land—on the immutable fact that a rising population will need to eat. But, of course, not everyone reads the footnotes.

It may be worth mentioning here that deforestation—the hacking of arable land out of the wilderness—is second only to fossil-fuel consumption in the list of economic activities causing global heating and climate breakdown. That said, Solow certainly knew nothing about climate change in 1956. And he may not have been seriously entertaining the possibility of the constant extraction of arable land: the slangy hyperbole of the verb "hack" in "as long as arable land can be hacked out of the wilderness" suggests he's lightly mocking the idea and, perhaps, himself. But later in his career he seems to take his ideas not as contributions to theory, but as explanations of the way the world really works. Collapsing the distinction between theory and practice, he began conflating his construct "technological possibilities" with a promise of technological substitution, and he started claiming that such substitution, done correctly, would allow humanity, in the real world, to transcend the limits the planet might place on growth.

Solow took this harder line in reaction to the publication of the 1972 bestseller *Limits to Growth*. Commissioned by the Club of Rome, this book was written by Donella Meadows and her group of systems scientists, Solow's colleagues at MIT. These scientists used still-nascent computer modeling to game out how economic growth and rising population might interact with planetary processes. Given the results of their model runs, these scientists warned that the world must wind down industrial production and maintain the global economy in a steady state or else, sometime in the twenty-first century, the Earth would run out of its "natural resources" and its capacity to absorb pollution, collapsing human systems and leading to the deaths of millions if not billions of people.

Limits to Growth was a sensation—translated into at least thirty languages, it sold millions of copies worldwide. Partly a product of the 1960s environmental movement—which resulted in the first Earth Day, the establishment of the Environmental Protection Agency, and the bipartisan passage of the Clean Air, Clean Water, and Endangered Species Acts—*Limits to Growth* also emerged out of the economic pessimism of the early 1970s, the era of growing "stagflation," low public confidence in the dollar after the abolition of the gold standard, and the 1973 OPEC oil embargo that quadrupled the price of crude. The news media had also primed the public for the message that ecological catastrophe was imminent. Although the slick of oil and chemicals in Ohio's Cuyahoga river, long used as a dump by midwestern factories, frequently caught fire in the mid-decades of the twentieth century, when the river ignited in 1969 the press generally reported the event as proof that industrial pollution was suddenly and completely out of control. At the same time, the press extensively covered the long lines of cars and irritated people waiting to fill up their tanks at gas stations, stoking fears of ongoing energy shortages and competition for scarce resources. In this media environment, the argument that the pursuit of economic growth would lead only to toxic pollution and resource depletion—and that these would lead to mass death—seemed remarkably plausible.

Professional economists, however, dismissed *Limits to Growth* with remarkable loathing. Reviewing the book in the *New York Times*, the Columbia economist Peter Passell called *Limits to Growth* "less than pseudoscience and little more than polemical fiction . . . best summarized not as a rediscovery of the laws of nature but as a rediscovery of the oldest maxim of computer science: garbage in, garbage out."[11] Writing with more patrician reticence but equal disapproval in *Foreign Affairs*, President Kennedy's former Deputy National Security Advisor Carl Kaysen found that "the authors' analysis is gravely deficient and many of their strongest and most striking conclusions unwarranted."[12] Like his colleagues, Solow disdainfully dismissed the book, calling it flat-out "bad science."[13] Yet he also took the more nuanced view that, yes, "the system might still burn itself out and collapse in finite time," but whether it does or not "all depends on the particular, detailed facts of modern economic life as well as on the economic policies that we and the rest of the world pursue."[14]

Of course, that was exactly what *Limits to Growth* itself had said, which is why it had called for a globally coordinated landing into a steady-state economy. But Solow insisted that this would be entirely the wrong approach. Instead, he called for policies that would encourage continued or even accelerated economic growth. He argued that growth would emerge from precisely the same innovations that would enable humanity to substitute technology for nature, and that growth itself, therefore, could solve environmental problems on the way to creating wealth. If this argument seems circular—to prevent limits to growth you encourage growth—that's because it is. But Solow apparently felt confident in making it for two reasons. For one thing, the models informing *Limits to Growth* did lack variables representing technological innovation or the mechanisms to encourage such innovation. They input only amplifying feedback loops—more production leading to larger populations leading to greater extraction leading to more production leading to more pollution, and so on—but no possibility that, as extraction and pollution increased, humanity might switch to less extractive and polluting technologies or processes (switching from fossil fuels to clean energy, for example). As Solow put it, "you hardly need a giant computer to tell you that a system with those behavior rules is going to bounce off its ceiling and collapse to a low level."[15]

The second reason Solow recommended more growth to solve limits to growth is that he shifted from only *theorizing* that "technological possibilities" would perpetually multiply productivity, to claiming that technological substitution was a principle that, in practice, could continue indefinitely in the real world. His shift was subtle but utterly crucial, for Solow now argued that technological substitution would both enable continued growth and allow humanity to transcend its planetary limits. The substitution of technology for "natural resources"—whether arable land, raw industrial materials, fossil fuels, or whole ecosystems themselves—would enable people in essence to use up the planet and still live healthy, safe, and affluent lives without fear of overshoot and collapse.

His position on substitution is perhaps best encapsulated by the summary he offered in his Richard T. Ely Lecture at the 1973 annual meeting of the American Economic Association. There, he announced, the "key factor" in the question of whether or not economic growth would lead to collapse is "the degree of substitutability":

> If it is very easy to substitute other factors for natural resources, then there is in principle no "problem." The world can, in effect, get along without natural resources, so exhaustion is just an event, not a catastrophe . . . If, on the other hand, real output per unit of resources is effectively bounded . . . then catastrophe is unavoidable . . . Fortunately what little evidence there is suggests that there is quite a lot of substitutability between exhaustible resources and renewable or reproducible resources, though this is an empirical question that could absorb a lot more work.[16]

If in his theoretical growth model he had substituted "technological possibilities" for land as a mathematical factor of production, here he claimed that humanity can actually substitute technological "reproducible resources" for elements of nature whose extraction has limits—"exhaustible resources," as he called them. To the extent of our capacity for substitution, economic growth could continue indefinitely, he maintained, giving us little need to worry about the exhaustion of "natural resources." He even entertained the possibility, as if it were a real possibility, that technology could almost entirely replace various elements of nature, enabling humans to "in effect, get along without natural resources." And, he argued, insofar as we want to enjoy nature, we can just substitute other resources for the ones we use up: "if you don't eat one species of fish, you can eat another species of fish," he remarked, perhaps a bit glibly, which "suggests that we do not owe to the future any particular thing."[17]

One might wonder, here, what Solow might have made of the potential extinction of not just a "species of fish," but the world's coral reefs, which provide the foundation for vast marine ecosystems as well as coastal protection, food, and livelihood to half a billion people. Perhaps he might have seen coral reefs as just another "amenity," a site for tourism and other forms of leisure, but not anything particularly essential to anybody. "I think that we ought to embody our desire for unspoiled nature as a component of well-being," he said in a 1991 lecture at the Woods Hole Oceanographic Institution, but at the same time, he added, "we have to recognize that different amenities really are, to some extent, substitutable for one another."[18] Perhaps Solow didn't notice that the phrases "really are" and "to some extent" pull in opposite directions—are amenities really always substitutable or not? And are ecosystems even "amenities"? In any event, Solow ultimately argued that the principle of universal substitution not only ensures ongoing economic growth but gives big consumers a moral free pass to use the planet as they see fit: "we know," he reassured his audience, "that one kind of input can be substituted for another in production," so "there is no reason for our society to feel guilty" about despoiling nature or using it up as long as we "leave behind a capacity to perform the same or analogous functions using other kinds of materials"—whether other "natural" materials, or "plastics," or other "artificial materials."[19] Yet it's not as if plastics or other artificial materials have no ecological costs. Plastics alone embody around 4.5 percent of annual greenhouse gas emissions.[20]

And what about the danger that "our society" might use up something irreplaceable? Wouldn't something irreplaceable be without substitute? For Solow, not at all: "When we use up something that is irreplaceable, whether it is minerals or a fish species or an environmental amenity, then we should be thinking about providing a substitute of equal value."[21] Paradoxically even the irreplaceable is somehow subject to exchange. Yet even at his most extreme in these talks, Solow was no ideologue. He also acknowledged, judiciously, that "if you approach the problem in that way in trying to make plans and make policies, it is certain that there will be mistakes"—and in the end he allowed that how much humanity really can substitute technologies and human-made materials for ecosystems is fundamentally not something you could know in advance, but, as we have seen, an "empirical question that could absorb a lot more work."[22]

Yet that empirical question was never really taken up by neoclassical economists, since they shared Solow's basic faith in technological substitution, and in any event their faith seemed to be repaid in abundance by the economic growth of the neoliberal era, which helped to make Solow's account the dogma of neoclassical economics. For decades the world seemed to work as Solow said it would. Over the 1980s and early 1990s in particular, economic productivity, including agricultural productivity, increased tremendously. The biologist Paul Ehrlich—then famous for his catastrophist book, *The Population Bomb*, which warned that hundreds of millions of people would starve to death in the 1970s and 1980s—lost a public bet with the libertarian economist Julian Simon, who wagered that the price of what he called "non-government-controlled raw materials" would drop in the long term because "innovators, working for profit or otherwise," would "seek out new lodes, develop new methods of extraction and invent new substitute ways of filling the same needs."[23] Ehrlich, by contrast, bet that scarcity would dominate and that, therefore, over the subsequent decade the prices of copper, chromium, nickel, tin, and tungsten would rise—but he lost his bet when they actually dropped. Ehrlich's loss of this wager—not to mention the inaccuracy of his Malthusian predictions for the 1970s and 1980s—helped to delegitimize economic work on limits to growth and enshrine the Solowian world view as the common-sense position.

Of course, had the Simon–Ehrlich wager ranged in term from 1900 to 1999, when the prices at issue remained largely stable, neither man would have won. Had it ranged from 1900 to 2008, when these prices rose, Ehrlich would have won.[24] But nobody was worried much about those bets by the time Simon's predictions were shown to be specious. (Like Simon, Solow himself made some dodgy predictions. In the 1970s he proclaimed that "no one can doubt" that "coal and nuclear fission will replace oil as the major sources of energy" and that "the production and consumption of oil will eventually dwindle toward zero."[25] But nobody worried much about *his* wrong-headed forecasts either.) Everyone was swept up in an enthusiasm for innovation. As Ronald Reagan once said, embodying the spirit of his age, "there are no such things as limits to growth, because there are no limits on the human capacity for intelligence, imagination, and wonder."

It's not entirely clear that the wealth of the neoliberal era, concentrated as it was in the top 1 percent, was really created by the glories of the human spirit. In some accounts it appears equally likely to have been manufactured by US policy. The economist Branko Milanovic has argued that the Volcker Shock of the early 1980s—when then Federal Reserve Chair Paul Volcker broke the stagflation of the 1970s with high interest rates that increased unemployment and thereby decreased the spending power of working people—"brought back confidence of the financial markets and stimulated large international capital inflows into the United States," showing foreign investors that the US government was ending the New Deal era and siding with capital over labor. This show of its willingness to break with labor, according to Milanovic, allowed the US government "to run forty years of uninterrupted current account deficits—a thing no other country in the world can dream of."[26] Decades-long deficits sustained supply-side fiscal policy, such as the reduction of the top marginal tax rate from 70 percent to 37 percent. And this in turn, as Thomas Piketty has argued, led to asset inflation misread as real economic growth, with the wealthy just parking their new-found liquidity in equities and other financial products. At the same time, legislators began to support profit-taking by deregulating business and finance and globalizing manufacturing. All these forces, as much as technological innovation, produced rising GDP, concentrated in the top 1 percent. More to the point here, whatever policies produced neoliberal economic growth, that growth added more greenhouse gases to the atmosphere between 1980 and 2020 than had been added in all previous human history.

Robert Solow died in 2023. Toward the end of his life, he seemed to become increasingly concerned about the climate crisis. When asked in his late interviews to identify a short list of pressing economic problems, he repeatedly placed climate change among the top threats to economic growth.[27] In a 2014 book entitled *In 100 Years: Leading Economists Predict the Future*, Solow warned that "even if we continue to think only of the already developed world" we have to acknowledge that "one hundred years is long enough for the effects of global warming to limit economic growth—perhaps marginally, but perhaps drastically."[28] In 2019 Solow and forty-four other economists including his fellow Nobel laureates Amartya Sen and Daniel Kahneman, along with former Chairs of the Federal Reserve Alan Greenspan and Janet Yellen, and former Treasury Secretary Lawrence Summers (although, curiously, not Nordhaus) together signed an open letter asking the US government to tackle the climate crisis by taxing carbon emissions. Arguing that it sends "a powerful price signal that harnesses the invisible hand of the marketplace," these economists asserted that a carbon tax "offers the most cost-effective lever to reduce carbon emissions at the scale and speed that is necessary" because by "substituting a price signal for cumbersome regulations" a price on carbon will "encourage technological innovation" and "promote economic growth."[29]

Of course, this letter makes essentially the same argument that Solow made in 1972, when he lambasted *Limits to Growth* for not including any price signals in its models, and when he called the price system "the main social institution evolved by capitalist economies" for "registering and reacting to relative scarcity."[30] In 1972, Solow believed a rising price would motivate the market to innovate and substitute new technologies for natural resources, thereby both promoting economic growth and overcoming any planetary limit to that growth. In 2019, Solow and his fellow economists apparently still believed that a price on carbon, instead of industrial policy or pollution regulation, will motivate the market to innovate, switch old technologies for new, and thereby produce growth and solve climate change at the same time. It is true, as we saw in the previous chapter, that the full decarbonization of the economy will raise everyone's real incomes from their savings on electricity, heat, transportation, and health care—and that the clean-electricity revolution, which will substitute technologies like solar and wind for fossil resources, will generate a burst of new economic growth and create millions of new jobs. Yet it's hard to see how a price on carbon would motivate even this one-time technological substitution of clean for dirty energy.

As the political scientists Leah Stokes and Matto Mildenberger have pointed out, carbon taxes are "politically toxic," because they ask voters to bear the costs of both climate change and its resolution in the face of what is still largely inelastic demand for fossil fuels. Forcing consumers to pay more for fossil energy when alternatives are not yet readily available is to virtually guarantee political backlash.[31] Even on a more theoretical level a price on carbon fails, because resolving the climate crisis requires not merely switching out one energy source for another, but a full systems change: the transformation of the ways we work, play, feed ourselves, move around, make things, buy things, throw things out, and so on. Insofar as the price system is supposed to manage supply and demand across the nexus of discrete natural resources, making those resources increasingly expensive as they near exhaustion in order to motivate innovation to replace them, it is hard to see how a carbon tax will encourage the replacement of an entire economic system that emits greenhouse gases from every pore. Under the paradigm of substitution, we would need to substitute in another whole economy.

Or another whole planet. The logic of substitution of technology for "natural resources" (insofar as a livable planet is merely a "natural resource" and not a miracle) reaches its apotheosis in the billionaires who call for humanity to leave Earth behind and colonize Mars. Elon Musk articulated just this logic when he once insisted on X, formerly Twitter, that we need to make life "multiplanetary" so we can "back up the biosphere, protecting all life as we know it from a calamity on Earth."[32] Musk's Frankensteinian fantasy that we can copy the entire biosphere to a planet lacking the capacity to support life is the *reductio ad absurdum* of the theory of economic growth by technological substitution. We cannot really bring the entire biosphere and all eight billion people alive today to Mars. Perhaps Musk meant he would make life multiplanetary by bringing *some* plants and animals and *some* people—presumably those who could pay for a seat on one of his rockets. If so, he exposed the disturbing underbelly of the belief that we can destroy our biosphere and livable climate—or, at least, the "climate niche" in which humanity appeared and thrived on this planet—and just jump to Mars: the people making that jump would be, as *Don't Look Up* dramatizes, only the wealthy themselves. The rest of humanity would be left behind on an incinerating Earth below.

But if it came to that, would there still be wealth? The idea that present affluence depends not at all on a stable climate system is absurd even within Solow's growth paradigm, which requires a rising tax on carbon to encourage the substitution of clean energy for fossil fuels. Today only around 22 percent of the global economy's emissions are taxed at all, and invariably at rates too low to motivate technology substitution, let alone wholesale systems change.[33] Far from being adequately taxed, fossil fuels are still supported by record subsidizes.[34] This means, of course, that climate change largely remains an externality to the economy—which is to say that its monetary and human costs are not included in the prices we pay for things—and *that* means that in some sense the vast majority of the world's wealth, indeed everything we see around us today, has been created while we are free-riding on the future, as if in a Ponzi scheme in which we are exhausting a livable climate, taking it from our children and calling it GDP. (Even just the word "externality" conjures up the fantasy that the economy and the planet are separate spheres.) In this growth paradigm, where humanity uses things up and just substitutes what it makes for what it has been given, it becomes inevitable that people will start imagining that we can substitute an engineered planet for Earth. To be sure, interplanetary substitution is a dystopian fantasy of salvation, although perhaps a fitting one in a historical period where our idea of happy immortality lies in imagining that our children are sure to be "better off" than we are, no matter how hot and deadly their planet might become. This kind of secular faith finds a fitting messiah in a billionaire who wants to colonize Mars. Such a man embodies the belief that wealth will surely save us.

Economic Growth and Adaptation to Climate Breakdown

By how much, and for how long, will economic growth continue if the world continues to burn fossil fuels and emit carbon dioxide into the atmosphere? There is surprisingly little research on this seemingly crucial topic. Economists who study the effects of global heating on growth rates are few and far between. As the Congressional Research Service puts it: "this field of study into the economic effects of climate change is relatively small compared to other types of economic or climate-related research. The relative dearth of studies makes it challenging to reach specific 'mainstream' conclusions about economic impacts."[35] The studies that do exist suggest that growth will continue apace. In a 2020 meta-analysis of the relationship between climate change and growth rates, the Congressional Budget Office (CBO) projected that climate change will contract US real GDP only 1 percent by 2050, relative to a world without climate change.[36] Of course, 1 percent is a mere cough of the market; 99 percent of the growth we might expect in an undamaged world we should also expect, according to the CBO, even as the planet gets hotter. This may seem like a very reassuring result. Yet this result holds for warming of only around 2°C, which climate scientists estimated in 2020 would most likely arrive around 2052.[37] If the world does not phase out fossil fuels and achieve net-zero emissions before then, the planet will continue to warm in 2053 and beyond.

It is also important to understand how the CBO economists achieved their result for 2°C of warming. Attesting to the lack of research on the question, they analyzed only four papers (rather than the typical dozens). And they not only correlated the findings of these four papers, but also "adjusted" their numbers using a function that assumed future adaptation to climate change. Deep in a footnote on their methods, they explained that only two of the four studies they compiled "explicitly modeled the potential effects of adaptation," and so "to better account for those effects, the CBO adjusted the historical relationships from the two studies that did not."[38] In other words, with some circularity, the CBO made climate change harm growth rates *less* by adding a function that represented their assumption that future adaptation will protect growth rates. This choice may recall the way Solow used "technological possibilities" as a production function in his growth model to set aside the limits that arable land places on population and productivity so as to assume ongoing constant rates of return. Here the CBO economists used "adaptation" as a function in their models to set aside the "historical relationships" between climate change and growth rates so as to assume ongoing constant returns. Doubtless governments, companies, and people will adapt to climate change—at least to some degree and at some cost—but for now, as it appears in economic models, "adaptation" is an adjustment that takes its value from the modeler's faith in the power of human ingenuity and our general capacity to solve problems with innovation.

Economists are very clear that the value of adaptation in their models is not grounded in any empirical evidence of what adaptation can or cannot do in the real world. In a study analyzing the long-term macroeconomic effects of climate change, a group of economists from the University of Southern California, the University of Cambridge, the International Monetary Fund, and National Taiwan University, who also use "adaptation" as a function in their models, explain, given the fact that even rich countries have not yet done much, if anything, to adapt to rising temperatures, their "analysis is counterfactual given the current state of the world." At the same time, they assert that "it is expected that adaptation weakens the relationship between temperature and economic growth over time."[39] These economists find the relationship between temperature and growth to be increasingly destructive in a hotter world: even including projected adaptation, they estimate in the worst case the United States would suffer a median loss of 6.81 percent GDP per capita by 2050 and 17.19 percent by 2100. In the better case, the United States would see losses as high as 3.77 percent and 10.52 percent at those dates. (For comparison, the 2007–09 Great Recession contracted US GDP by 5.1 percent per capita and the Great Depression by about 30 percent.) These numbers—which are underestimates insofar as these economists' beliefs about adaptation are too optimistic—are still more alarming than those produced by the economists at the CBO. Yet this study is included in the CBO's meta-analysis, which suggests just how far the CBO "adjusted" its results using "adaptation" as a function in its models.

The belief that adaptation will enable continued GDP growth even as the planet heats up and the climate breaks down sometimes gets extended into the more extreme belief that adaptation will enable economic growth to continue even if the world never stops using fossil fuels. Resident *Wall Street Journal* lukewarmer Holman Jenkins Jr., whom we first met in "Alarmist," often professes this very belief, writing in one commentary, for example: "It would be especially expensive to adapt to an especially dire climate scenario. No kidding, Sherlock. But climate varies and humans do adapt. Populations and economic activities (like farming) relocate over time. Coastal communities pull back or build dikes. This is costly, but not so costly that average human welfare won't keep going up, up, up."[40] Jenkins is a fossil-fuel apologist, of course, writing column after column attacking climate alarm, renewable energy, and the Democrats' climate policies. But his view has been echoed by Michael Barbaro, the avowedly climate-conscious host of the *New York Times* podcast "The Daily," who once said on air that "the worst impacts of climate inaction are probably never going to be felt by places like the US and the countries of Europe . . . mostly wealthy countries that can afford to adapt to climate change."[41] Really, never? People in wealthy countries are already being killed by extreme weather. According to researchers, the 2022 heat wave in the United Kingdom and Europe—to which Barbaro noted the United Kingdom was adapting by, for example, "wrapping [their] bridges in foil"—killed nearly 63,000 people.[42] Of course, it's also true that those deaths are not counted in GDP.

Presumably some of the people killed by the heat wave would have lived had they used air conditioning. For its power to save lives air conditioning is a key adaptation to climate change. Yet indoor cooling cannot offset all the mortal harms of global heating, especially upon those who cannot afford to buy an a/c unit or who work outdoors. Nor can air conditioning mitigate all the economic harms of global heating, even in industrialized nations. International-finance researchers have demonstrated that despite the adoption of adaptive technologies like air conditioning, "summer temperatures have a pervasive effect in the entire cross section of industries, above and beyond the sectors that are traditionally deemed as vulnerable to changing climatic conditions."[43] In the decades since 1990, when temperatures remained above 90°F for six or more days in a row, production at automobile manufacturing plants, for instance, dropped by an average of 8 percent. "Given that automobile manufacturing largely takes place indoors," these researchers note, "this finding suggests that there are limitations of air conditioning; it is possible that there are important areas in the production process, such as loading and unloading areas, which are difficult to cool."[44] The dawning realization that substituting industrial for agricultural production, moving labor and capital indoors, may not entirely protect the economy from the climate crisis undermines the Solowian assumption that human beings can always substitute the manufactured for the "natural" and thereby build artificial environments that are somehow separate from the planet on which they exist.

Yet this assumption has directly influenced climate economics. In 1993 Nordhaus argued that countries like the United States are "relatively insulated from climate change," because sectors that have "little direct interaction with climate," such as "underground mining, most services, communications, and manufacturing," comprise "around 85 percent of GDP." Nordhaus was especially confident about the growth potential of the most technologically advanced sectors, such as "cardiovascular surgery and parallel supercomputing," since these, he said, are undertaken in "controlled environments."[45] Twenty-five years later, in his 2018 Nobel Prize lecture, he expressed the same confidence, claiming that "many sectors of high income countries (such as laboratory clean rooms) are highly managed, and this feature will allow these sectors to adapt to climate change at relatively low cost for at least a few decades."[46] But just four years later, the 2022 heat wave shut down Google and Oracle data centers in the United Kingdom, interrupting a range of cloud-computing services, after the cooling infrastructure in the centers failed.[47] The same heat wave suspended operations in factories across the Sichuan province of China after authorities were forced to reserve electricity for household air conditioning.[48] Even some Chinese microprocessor factories needed to close. These industrial shutdowns gummed up the supply chains for semiconductors, photovoltaic cells, iPhones, and Teslas, whose production was thereby curtailed by warming of just 1.2°C.

This real-world event illustrates some provocative research into the relationship between temperature and growth by Marshall Burke and Solomon Hsiang, the Stanford and Berkeley economists we met in "Cost," who calculated the vast economic benefits of halting warming at 1.5°C. Here Burke and Hsiang, with their co-author Edward Miguel, have found that a country's economic productivity is at least partly determined by its relationship to a "goldilocks" temperature range, and within the framework of this relationship, the economies of rich and poor countries respond virtually identically to warming. Using data tracking the interactions between temperature and each nation's economic productivity since 1960, Burke and Hsiang identify an optimum average temperature for GDP growth—around 13°C or 55.4°F—around which the vast majority of global economic production occurs. (Europe and the United Kingdom are very slightly cooler than this optimum; the United States, Japan, and China are slightly warmer. Most other countries are much warmer.) As they put it: "Cold-country productivity increases as annual temperature increases, until the optimum. Productivity declines gradually with further warming, and this decline accelerates at higher temperatures."[49] And this accelerated decline in productivity at higher temperatures is, they show, "globally generalizable, unchanged since 1960, and apparent for agricultural and non-agricultural activity in both rich and poor countries."[50] If global heating has not yet significantly contracted the economic productivity of rich, industrialized nations like the United States, that does not mean, according to this research, that these nations respond differently to warming; it means that these nations are still relatively close to the temperature optimum.

This is certainly a counterintuitive discovery. It seems incontrovertible that wealth shields people from climate change (just compare riding out a cyclone in a hut of tin and straw to sheltering in a building of steel and concrete). But Burke and Hsiang see rich and poor countries' identical response to temperature in the historical evidence, finding that the economic response of rich countries to warming is "statistically indistinguishable from poor countries at all temperatures."[51] Burke replicated these findings in a 2019 study that examined the relationship between temperature and GDP growth in 380 US cities, all in the top 20 percent of global incomes, and found that "in this sample of very wealthy cities, the estimated response function is . . . nearly indistinguishable from the pooled global response."[52] There is "no evidence," in short, that "technological advances since 1960 have altered the global response to temperature." This lack of evidence that technology protects economies from the effects of temperature implies, according to Burke and Hsiang, that an "accumulation of wealth, technology and experience might not substantially mitigate global economic losses during this century," and, moreover, that "adaptation to climatic change may be more difficult than previously believed."[53]

Whether adaptation is difficult or not, the very need for adaptation to climate change belies the myth that the economy and the planet are separate spheres. If they were truly separate, then human beings would not need to adapt to ecological realities. Climate change reveals the porousness of the boundaries between the "insides" of concrete factory buildings, where productivity drops when it is extremely hot outdoors, and the "outside" of the so-called "natural world," exposing the limits of adaptive technologies like air conditioning that were thought to offer exquisite control over the built environment. Climate change blurs Robert Solow's world picture of the global economy as an ever-growing container of prosperity that extracts "natural resources" from and emits waste into a separate "nature." And, finally, climate change destabilizes the myth of "nature" as a nice-to-have amenity or, conversely, an "externality" whose limits (represented as costs to the price system) are sure to be surmounted by innovation, whether instigated by markets or by industrial policy.

Neoclassical economists are, however, still modeling adaptation as a kind of innovation, using it, as Solow used "technological possibilities," to ignore ecological realities and adjust their results to show perpetual growth. Yet adaptation doesn't always lead to growth. Sea levels are projected to rise around two feet by 2100, no matter what the world does now, and for this reason people and businesses will need to retreat from the coastline.[54] This retreat is absolutely a form of adaptation, but it could also contract insurance markets and potentially strand billions in tangible assets. On the world's current emissions trajectory, roughly two billion people may need to migrate away from regions that global heating could make uninhabitable as soon as 2030.[55] These people too would be adapting to the climate crisis—chasing cooler ecosystems by moving toward the poles is how most creatures will adapt to rising temperatures—but it is hard to imagine how the geopolitical instability produced by billions on the move might sustain economic growth.[56] Yet move people must, and be welcomed where they need to go. When land no longer produces enough food, when it is too hot and dry (or wet) to farm, adaptation becomes no growth function in a model, but a matter of life or death.

On their part, economists acknowledge that arable land is highly exposed to climate change. But because they see agriculture as a mere "sector" of global GDP—and a minor sector at that—they seem to believe that whatever damage climate change may do to agriculture will have only minor economic effects. For example, in one paper on climate change, adaptation, and growth rates, the Duke economist Richard Newell reassures his readers that climate change will hardly reduce growth rates at all, given the small "overall historical exposure of the economy to temperature," because agriculture "represents only a few percent of global income, whereas about two-thirds is services."[57] (Newell is a member of the National Petroleum Council—but also a former Administrator of the US Energy Information Administration under Obama.) The economists at the Congressional Research Service also see agriculture as a minor "sector," noting in one report that "climate change could cause productivity losses *in this sector*," resulting only in "*slower sector growth* than would be realized without climate change." These government economists also caution that reduced agricultural productivity means that "jobs could be lost"—although they also invoke the principle of substitution by noting that "workers would likely relocate to different regions or take jobs in different industries."[58] Their suggestion that declining agricultural yields would be a problem that the labor market could solve with substitution speaks to the distorted frame with which neoclassical economists see the production of food: as a small percentage of GDP—a sideline on the playing field where capital, labor, and innovation generate growth—rather than the foundation of human systems.

It is not entirely surprising that economists appear relatively sanguine about the cascading, systemic effects of declining agricultural productivity. Their theory of growth does not include land as a factor of production. And, truth be told, agriculture no longer appears subject to limits. With productivity having risen steadily over the twentieth and twenty-first centuries, especially since 1960, famines now seem preventable, to distort Amartya Sen's famous dictum, simply by the exercise of democracy. The world produces so much food that 25 to 30 percent of it goes to waste every year.[59] At the same time there remain deep injustices in global systems of food commodification and distribution. Millions of people in Africa, including an estimated sixty million African children, still suffer from hunger and malnutrition.[60] Even some Americans are undernourished, despite eating a surplus of calories, because they live in what sociologists call "food deserts," where stores sell only gas-station fare, highly processed packaged meals and snacks loaded with preservatives, artificial flavors and colors, salt, and corn syrup. But even given the horrific failure of markets to allocate food efficiently and equitably, it remains the case that agricultural productivity is as high as it's ever been.

That may no longer be true if the planet heats up to 3°C of warming. In its 2022 report on "Impacts, Adaptation and Vulnerability," the Intergovernmental Panel on Climate Change (IPCC) warned that "heat and drought are projected to substantially reduce agricultural production," and that climate change will harm food systems and human health through "simultaneous reductions in food production across crops, livestock and fisheries." It also notes that "although irrigation can reduce this risk, its feasibility is limited by drought."[61] In its Special Report on Climate Change and Land, the IPCC projected that on an emissions trajectory leading to 2.8°C of warming by 2100—our current emissions trajectory, as we saw in "Alarmist"—the price of cereals will rise up to 29 percent by 2050.[62] Such a striking rise in food prices will certainly put pressure on consumer spending throughout the economy, retarding growth while leading to sharp increases in the number of people who don't have enough to eat.

Yet this estimate of how food prices may rise, worrisome as it may be, is actually an underestimate of the ways that climate change could damage agriculture and the economy. In a 2022 analysis, food-systems scientists at Harvard, Johns Hopkins, and the International Food Policy Research Institute (IFPRI) cautioned that "projected effects of climate change on food security are often based on crop models that incorporate only a few dimensions of climate related biophysical change"—usually just temperature and precipitation—while omitting "other biophysical changes related to a disrupted climate system" that are harder to represent quantitatively.[63] So even when they incorporate drought into their projections, crop models will leave out the converse—the destruction of crops due to torrential rain and flooding. They will also set aside the more subtle but no less important systemic effects of a warming planet, such as reductions in the nutritional content of staple grains due to rising CO_2 levels, the increase in the range and virulence of crop pests and livestock pathogens, delayed "chill accumulation" that causes fruit-set failure, too-early spring warming that encourages fruit flowering which is then blighted by traditionally timed frosts, high spring temperatures that kill pollens and prevent crop fertilization, mismatches between animal pollinator behavior and shifting planting seasons, and the migration of fish north away from the marine territories where we have built industries to catch them. These are the more systemic effects of climate change on our food systems, in excess of the effects of heat and drought alone. And as the researchers from the IFPRI, Johns Hopkins, and Harvard put it, "these effects are difficult to quantitatively represent because our level of understanding of the underlying dynamics is limited."[64]

It is perhaps alarming enough to consider only the effects of heat and drought. Using just these metrics, scientists have identified worrisome signals that climate change is already harming the global food system. Scientists at Cornell University recently discovered that the amount of food produced per acre of land has declined 21 percent globally between 1961 and 2021, with 26–34 percent declines in Africa, Latin America, and the Caribbean.[65] In 2022, the Food and Agriculture Organization of the United Nations announced that millions of people were on the verge of dying from starvation in East Africa after what they called an "unprecedented" drought killed millions of cows and withered crops in what turned out to be five failed rainy seasons in a row.[66] In Africa's Sahel region, rising food prices due to speculation and Russia's invasion of Ukraine intersected with torrential rainfall, which led to another unprecedented event: a flood in Chad that destroyed nearly 2,700 hectares of crops and farm land, threatening millions more with death by famine.[67] Scientists found that this flood was made eighty times more likely by climate change.[68]

In north Africa and the Middle East, the situation was hardly better. Between 2021 and 2022 all rain-fed and 50 percent of irrigated wheat production in Syria failed due to drought.[69] In Turkey, wheat production dropped 13.9 percent.[70] Meanwhile in the Indian states of Punjab, Haryana, and Uttar Pradesh, together home to about 286 million people and a core producing region for the global food system, the spring 2022 heat wave resulted in an estimated 10–35 percent reduction in crop yields.[71] Right next door, the catastrophic floods in Pakistan destroyed 65 percent of the country's rice and wheat fields.[72] Floods also destroyed 6.7 percent of the cropland in the Henan province of China,[73] and the unprecedented late-summer heat wave, in which record-breaking temperatures baked the entire nation for over two months straight, forced farmers to begin their work in the middle of the night, when it's coolest, although this adaptation didn't end up saving many crops. "The key is it's not possible to grow them," one farmer, Wang Yong Fu, told a reporter, "whatever you do, you can't grow anything . . . the crops have all died."[74]

Europe's farmers also struggled with this infernal trifecta of flood, drought, and heat. Weeks of extreme temperatures reduced the 2022 French corn harvest by 25 percent.[75] By the end of that summer, a third of Italy's agricultural region withered under a declared state of drought emergency.[76] In the United Kingdom, half the potato crop was expected to fail, and shoppers were told they needed to learn to accept undersized produce after emptying reservoirs prevented farmers from being able to fully irrigate their fields.[77]

Agriculture has been struggling in the United States too. In 2022 American red winter wheat produced its lowest yield since 1963, rice was down by half, and crops like spring wheat, durum wheat, and barley saw the smallest harvests since the 1980s. The tomato harvest was nearly four million tons shy of what it was in 2015; the weight of the potato harvest dropped 4 percent from 2021 (then already the lightest harvest in a decade); and in September shipments of carrots were down by 45 percent.[78] The growing regions of the central plains—Kansas, Oklahoma, Nebraska, the Dakotas—were particularly hard hit by a full range of extreme weather, from floods in North Dakota that drowned up to 30 percent of one farmer's fields to heat waves that killed at least 2,000 cows in Kansas in June alone.[79]

California currently grows about half of US produce, and it's home to some of the largest dairies, pork feedlots, and cattle ranches in the country. The San Joaquin Valley by itself produces nearly half the state's agricultural output. But the region's productivity is threatened by what appears to be California's new climate, in which extended droughts are punctuated by extreme precipitation. In 2022, for example, nearly all of California suffered from "extreme drought" or worse, with over 10 percent of the state in "exceptional drought," the most severe category, that May through September.[80] Starting that December, however, multiple atmospheric rivers flowed in from the Pacific and dumped so much precipitation on the state that most of its counties were covered by emergency flood proclamations by the end of March.[81] Yet even then California's surface water reserves remained low—by October 2023 Lake Mead and Lake Powell storage remained at 38 and 34 percent of average, respectively—and the state's groundwater remained significantly depleted.[82] California farmers have drawn so much water from the ground that the land in the Central Valley is sinking—in some spots more than a foot a year.[83] To try to hold off the depletion of its freshwater reserves, the state recently passed the Sustainable Groundwater Management Act, which requires that the use and replenishment of the state's groundwater come into balance by 2040. To achieve that balance, farmers in the Central Valley would need to reduce their groundwater pumping by 70–80 percent compared to 2022. One farmer, who helps manage 40,000 acres near Fresno, told a reporter that he's found groundwater irrigation to be "a lifeline that we've all taken for granted," but now he and his fellow farmers are realizing that "it's not infinite."[84]

The water situation in California—historically bad, now made worse by the climate crisis—really brings home the IPCC's warning that as an adaptation, irrigation can reduce the risk of declining yields, but "its feasibility is limited by drought." In adapting to drought years, some California farmers are buying lightly filtered oil-drilling wastewater from fossil-fuel companies. About 100,000 acres of nuts, citrus, and vegetables, even those certified organic, are irrigated with this chemical-laden slurry.[85] Setting aside oil-drilling wastewater for being a potentially hazardous source of irrigation, farmers might turn to desalinated seawater as another agricultural adaptation. Desalinating seawater is energy- and cost-intensive, however, and it may not be realistic to pipe billions of gallons of treated water from the Pacific to California's Central Valley, hundreds of miles away and itself 18,000 square miles in area.

Another potential adaptation would be to build more distribution infrastructure for the agricultural regions of the Pacific Northwest and the Canadian prairie, which will likely become more productive as their growing season lengthens and their climates warm. Yet this adaptation too presents problems. Although global heating has in theory expanded the agricultural potential of northern latitudes, it has also altered the growth stages of wheat, potentially causing plant-pollinator mismatches in crops that farmers might attempt to establish in new places. As ecosystems move north, pests will begin to thrive in upper Canada, Russia, and Scandinavia, with projected damages to wheat yields from pests alone increasing by 10–25 percent for every degree of heating.[86] And it would be a poor trade-off to develop agriculture in northern regions by cutting down boreal forests—or by hacking arable land out of the wilderness, to paraphrase Solow—because this deforestation would release carbon into the atmosphere and destroy important natural carbon sinks. Even as the world needs to feed a growing population, it also needs urgently to address the greenhouse gas emissions of agricultural production, currently 10 percent of the global total, and radically reduce the ecological footprint of farming, currently the greatest contributor to deforestation and the extinction of biodiversity.[87] (Phasing down animal farming, and allowing ecosystems to repopulate bare pastureland, would go a long way to achieve this goal: cows, pigs, and other livestock now use nearly 80 percent of all agricultural land, while producing less than 20 percent of all calories.)[88] The need to transform food production so that it stops the destruction of our livable climate will be in conflict with the drive to adapt to diminishing agricultural productivity by extracting new arable land out of forests in northern latitudes. Here too "adaptation" quickly comes up against its limits.

It is, certainly, extremely challenging to quantify the relationship between diminishing agricultural productivity and economic growth rates. But in a 2020 study, the management consultancy McKinsey & Company tried to envision what this relationship might look like. First they estimated that a true multiple-breadbasket failure—simultaneous "acute climate events" in a sufficient number of breadbaskets to significantly reduce global grain reserves—could "easily" cause global food prices to "spike by 100 percent or more in the short term." Then they noted that "negative economic shocks of this size could lead to widespread social and political unrest, global conflict, and increased terrorism"—all phenomena that tend to contract global markets. Finally, they set the likelihood of such a shock happening "at least once in the decade centered on 2030" at 18 percent—worse odds than Russian roulette.[89] (A 2021 study partially funded by NASA also found that the major breadbasket regions of our planet will be significantly harmed by global heating "before 2040.")[90] Now, the global economy can recover from and even grow after wars and other kinds of temporary shocks, but the question is how often "acute climate events" in the world's breadbaskets—California, Brazil, the US Upper Midwest, the Canadian prairie, northwestern Europe, northern India, eastern China—will begin to occur if the planet gets hotter and hotter. Every summer? Every other summer? In any case, if climate change causes multiple-breadbasket failure, and this failure leads to "widespread social and political unrest, global conflict, and increased terrorism," perhaps economists should stop talking about agriculture as being less important to the economy than, say, financial services. Perhaps the foundational importance of agriculture for the stability of human systems is not reflected in its small share of global GDP.

The importance of agriculture to stability, and thus to growth, will only increase as the climate breaks down. Insofar as repeated heat waves and droughts make arable land increasingly scarce, to put it in Solowian terms, then climate change also challenges the myth that "technological possibilities" can ensure endless growth. As we have seen, even Solow acknowledges that endless growth is possible only "as long as arable land can be hacked out of the wilderness at essentially constant cost." Fossil-energy interests will insist that human beings can, in fact, hack arable land out of the wilderness at constant cost, easily relocating planetary breadbaskets to Russia and Canada—or they will say that scientists will innovate new technologies that will enable farmers to adapt to rising heat and deepening drought. But these adaptations, if they are possible, will not be easy. And, as Solow put it, "if it is very easy to substitute other factors for natural resources," only then (and only "in principle") is there "no 'problem.'" Here, in our world gripped by the fossil-fuel system and getting ever hotter, there is a problem, and economic growth is not the solution.

How to Talk about Climate Change and Economic Growth

The effects of climate change on arable land should not only attenuate the power of Solow's theory of growth, but also challenge the way climate communicators talk about the relationship between climate change and economic growth as a whole. Instead of repeating the received wisdom that growth will protect affluent people from the worst effects of climate change, no matter how hot the planet may get, it is important to convey that if it goes unchecked climate change will undermine the coupled human and ecological systems, like agriculture, that enable economic growth in the first place. Disenchanting the myth that growth is something that will protect the world from climate breakdown, you should make it clear that the world must halt global heating if it's going to protect the prosperity of the next generation—of our very own children. At the very least, you should use the latest economics research and current mortality statistics to emphasize that as climate change accelerates, it will hurt and kill ever larger numbers of the affluent, as it's already hurting and killing so many of the poor.

At the same time your message must absolutely call for climate justice. It is morally wrong and repellent to talk about any danger to the affluent without centering the fact that their prosperity was built on relations of colonial domination—in which, with the assistance of sovereign states and their militaries, and in the name of growth, Global North capital forcibly opened markets, enslaved people, extracted resources, and left behind its waste and its harms—including the increasingly devastating harms of greenhouse gas emissions—in the poorer regions of the globe. And it is not just morally wrong, but also politically ineffective to talk about climate action without talking about climate justice, because the unstated claim that some places, some people, and, in some ways, the planet itself are all dispensable only perpetuates the myth that says we can use up so-called "natural resources" as long as there are technologies to replace them. This myth must be superseded by new economic thinking that will help halt global heating, not just by modeling the substitution of zero-carbon for fossil energy, but also by contributing to the transformation of all the systems, each intertwined with the production and consumption of energy, that also destroy biodiversity while adding greenhouse gases to the atmosphere: industry, commerce, transportation, cities and suburbs, and, indeed, agriculture itself.

In order to halt global heating and enable degraded ecosystems to heal, the economy, science tells us, cannot extract more resources than the planet can regenerate nor emit more pollution than it can absorb. The economy must achieve balanced integration with the biosphere. Of course human beings will create new, transformative technologies to be intertwined with planetary processes as we transform the global economy—as, for example, vaccines interact with the body's own immune system, priming it to do its biological work—but we should renounce the tragic illusion that artificial materials can somehow replace what is known as nature. For, in truth, our so-called "artificial materials" *are* nature. Our plastics are nature. Our skyscrapers are nature. Our airplanes are nature. Our oil wells are nature. Everything made by human beings emerges through the ecology of this planet and remains in its systems—even the satellites that will one day drop back down to Earth, pulled inexorably by the gravity of home. This is the lesson climate change is trying to teach us; the only question is whether you and your fellow climate communicators can make the powerful see and learn that lesson in time—or help enough people absorb the lesson and come together collectively to disempower the people willing to ignore it.

If it seems like it would be impossible to help people talk about climate change and economic growth in a new way, consider: Robert Solow himself appears to have changed his mind. Not only did he warn, as we have seen, that "even if we continue to think only of the already developed world," climate change might "limit economic growth—perhaps marginally, but perhaps drastically." He also apparently lost his faith in technological substitution as the growth-producing solution for ecological crises, observing in an interview towards the end of his life that "the continued and expanded use of nonrenewable natural resources could lead to either effective exhaustion or sharply rising relative prices, either of which would alter growth prospects over a century." Even more remarkably, in the same interview Solow posited a regenerative, steady-state economy as the solution to the problem of extraction and pollution: "if the world were to go Voltairean and choose mainly to cultivate gardens," Solow proposed, "the input-output table could change a lot. Production would put less strain on the scarce resources base and the waste disposal capacity of the environment."[91]

This is a stunning reversal, from joking about hacking arable land out of the wilderness to musing about a cyclical economy, represented by the creative yet bounded practice of gardening, in one lifetime. Of course, it's hard to know what to make of this reversal—it's almost like an improbable ending in one of Shakespeare's late plays where the child who was thought to be lost ends up found—but the very least we can say is that Solow can stand as an example and a beacon of hope. If even Solow can reevaluate his life's work, rethink his deepest assumptions, and come to a new understanding, surely many others can see beyond the myth of perpetual growth—or at least be encouraged to try. (Solow's invocation of Voltaire's *Candide* is especially apt here, since that novel's titular hero decides to settle down and cultivate his garden only after enduring horrific trials to which he has been subjected by his relentless Panglossian optimism. Yet by accepting life's limits, Candide finds happiness in the end, even though that happiness took a different form than he had imagined.) You can encourage people to accept planetary limits by helping them reevaluate the unexamined beliefs that decades of Solowian economics have made conventional wisdom. And in helping people start talking about how climate change threatens the world's prosperity, you can help them see what is so important to know now: that future progress relies not on our so-called mastery of nature, but on our acceptance that we *are* nature, and in destroying nature we destroy ourselves.

4

"India and China"

Words matter in Chinese politics.

—Odd Arne Westad, Yale University
("What Does the West Really Know About Xi's China?")

China will be carbon neutral, and it will have achieved fruitful results in ecological civilization and reached a new level of harmony between humanity and nature.

—The Communist Party of China
("Working Guidance for Carbon Dioxide Peaking and Carbon Neutrality in Full and Faithful Implementation of the New Development Philosophy")

It is precisely because the [Chinese] Communist Party regime is bent on shaping the next century that its leader takes climate change seriously.

—Adam Tooze, Columbia University ("Welcome to the Final Battle")

Fossil-fuel interests often circulate the misleading talking point that so long as "India and China" continue to use fossil fuels, it won't matter if the United States tries to halt global heating—any American effort will be futile. This talking point is designed to deflect the United States' historical responsibility for climate change and to sow hopelessness and cynicism among the electorate. The *Wall Street Journal* deploys this obstructionist tactic vigorously and often. Noting in one commentary that the United States has pledged to zero out its emissions by 2050, but hasn't yet "implemented the policies or developed the technologies to get there," the Editorial Board goes on to say that enacting robust climate policy in the United States would not "make much difference . . . as long as China and India continue to build coal plants to fuel their economic growth."[1] In another piece, the Board warns that people in India and China are running headlong into climate destruction and taking the world with them: "Anything the US does to reduce emissions won't matter much to global temperatures," they insist, for "US cuts will be swamped by the increases in India, Africa, and especially China."[2] In yet a third piece, the Editors leaven their justification for embracing fossil energy by adding a little geopolitical fear to the mix, implying that phasing out fossil energy is not only futile, but self-destructive, because only widespread oil and gas use ensures America's continued global hegemony: "While the Biden Administration does all it can to restrict U.S. fossil fuels, no matter the economic harm, Beijing is charging ahead with coal imports, coal mining, and coal power to become the world's leading economy. They must marvel at their good fortune in having rivals who are so self-destructive."[3]

Highlighting the fossil-fuel development of "India and China" has been very effective. When asked who is to blame for global warming, fully 68 percent of registered voters point to "developing countries"—more than the number of voters who blame fossil-fuel companies or even industrialized nations, the true source of most fossil-fuel emissions to date.[4] But "India and China" propaganda preys on feelings of helplessness that plague even people on the left. The food writer Michael Pollan, for instance, asks in his one essay on climate (entitled "Why Bother?") what the point would be of his doing anything about climate change when he knows "full well" that, as he says, "halfway around the world there lives my evil twin, some carbon-footprint doppelgänger in Shanghai or Chongqing . . . who's positively itching to replace every last pound of CO2 I'm struggling no longer to emit."[5] It's striking that even such an observant writer as Pollan gets the villain so wrong: not governments supporting fossil-energy with their policies, nor fossil-fuel companies fighting the transition to a safe-energy system, but a Chinese "evil twin" who's "positively itching" to emit carbon dioxide. Hiding the true culprits behind a vaguely racist caricature of an Asian "evil twin" is exactly the cognitive effect the "India and China" tactic is designed to achieve.

It is certainly true that emissions in India and China are rising—as they are across the developing world—and that these emissions must fall to net zero in order to halt global heating. But "India and China" is not a coherent political entity wholly committed to fossil-fuel development. India, for one, has set a target of net-zero emissions, and whether it achieves this target depends a great deal, I will show, on the support it receives from the United States and the rest of the Global North. Even as India has nuclear weapons, a space program, and a rising middle class, it remains a lower middle-income country with the largest population of poor people of any nation in the world.[6] Modernizing India's infrastructure, expanding its productive capacity, and strengthening its economy without entrenching fossil-fuel development will require new finance and technology transfers from the wealthy world.

China too has a net-zero target. More than that, I will reveal, it has a claim to climate leadership that surpasses that of the United States. China has become the world's foremost producer and distributor of clean-energy technology—and it is deploying that technology at home with astonishing speed. Even more startling: China has developed an elaborate policy architecture to decarbonize its economy in the shortest period of any nation on earth, and it's taking concrete steps to make sure those policies are fully implemented. Rather than being the villain, China appears to be setting itself up to become the climate hero of the twenty-first century.

This message belies the propaganda of fossil-fuel interests, of course, but it also complicates the mainstream narrative that—at least under Democratic presidents—America is the world's climate leader whose diplomacy has been thwarted by lack of cooperation from authoritarian petrostates and "India and China." In truth, as I will reveal, even under Democratic administrations the United States has undermined the structure of global climate governance and consistently blocked international climate agreements from including legally binding targets. As the largest historical emitter and the world's wealthiest nation, the United States has always borne the greatest responsibility for the climate crisis, but with the full consent of the Democratic party it has consistently shirked its duty to the world community, giving China the opportunity to corner clean-energy markets, develop new policies to fight global heating, and one day perhaps even set the norms of international climate governance.

It is not too late for the United States to become the climate leader of the twenty-first century. But it must delay no longer. If China makes obvious headway decarbonizing its economy while progress in the United States continues to drag, the cultural supremacy of American democracy will begin to wane. Conversely, if China never fully implements its climate policies, only the United States, in concert with its allies, will retain enough geopolitical power to deploy cross-border measures like climate clubs and sanctions at a scale that can force China to fulfill its promises.

Fossil-fuel interests and some on the left say that American climate action is futile because "India and China" will never achieve net-zero emissions. To fight this propaganda, here is a new message: under both Republican and Democratic administrations, the United States has so far continued to expand fossil-fuel production, while China has positioned itself to dominate the global transformation to a net-zero economy. To sustain its preeminence, America must make and uphold binding international climate commitments to phase out fossil fuels and create an ecologically integrated global economy. And Americans can demand this! So long as the United States remains a democracy that guards the rights of free speech and assembly, its citizens can compel their elected representatives to act in their interests on the world stage.

America Has Been the Problem

The structure of international climate politics was established by the United Nations Framework Convention on Climate Change (UNFCCC), the 1992 treaty ratified by 198 countries that established the regime through which nations were to cooperate "to prevent dangerous anthropogenic interference with the climate system." The first principle of the UNFCCC is that nations "should protect the climate system for the benefit of present and future generations of humankind, on the basis of equity and in accordance with their common but differentiated responsibilities and respective capabilities."[7] This means that every nation shares the responsibility to protect the climate system, but the extent of that responsibility is apportioned not equally, but on "the basis of equity," or according to the degree a nation has already damaged the climate with its historical emissions. It also means that what nations are expected to do under the norms of the climate regime will depend on their "respective capabilities," or their level of economic, technological, and institutional development. In other words, as its first principle, the UNFCCC set up staggered starting blocks for the race to what is now understood to be a net-zero emissions, ecologically-integrated global economy, allowing developing nations to continue to use fossil fuels in order to grow their gross domestic products (GDPs) and eradicate poverty. As the Convention reiterates, all nations "will take fully into account that economic and social development and poverty eradication are the first and overriding priorities of the developing country Parties."[8]

Accordingly the UNFCCC mandated that "the developed country Parties should take the lead in combating climate change and the adverse effects thereof." To "take the lead," here, means two things at once. Primarily, it means being the first to phase out fossil fuels in order to make space in the global carbon budget for, by way of example, infrastructure development using steel and cement, whose zero-carbon forms are still being innovated. Additionally, to "take the lead" means using some of the wealth that fossil-energy development has created in the Global North to help finance development that leapfrogs over fossil fuels directly into sustainable systems. The UNFCCC, with the binding word "shall" rather than the purely recommendatory "should," enshrined the obligation that developed countries "shall provide new and additional financial resources to meet the agreed full costs incurred by developing country Parties in complying with their obligations . . . including for the transfer of technology."[9] Indeed, the treaty put the responsibility for the success of the global climate regime directly on the shoulders of wealthy nations: "The extent to which developing country Parties will effectively implement their commitments under the Convention will depend on the effective implementation by developed country Parties of their commitments under the Convention related to financial resources and transfer of technology."[10]

All international climate commitments—from the 1997 Kyoto Protocol, through the 2009 Copenhagen Accord, to the 2015 Paris Agreement—remain guided by the UNFCCC principles of "equity and common but differentiated responsibilities and respective capabilities, in the light of different national circumstances," as the Preamble to the Paris Agreement puts it.[11] Yet the United States has consistently violated those principles both in its negotiations of these agreements and in its subsequent behavior. Not only has the United States ensured that every nation is equally obligated to reduce emissions, no matter how poor, it has also refused to support legally binding climate targets and invariably reneged on its promises to provide finance to the Global South. Yet US government spokespeople as well as the American news media have largely blamed India and China for blocking a legally binding treaty to phase out fossil fuels. For decades, under the principle of equity, India and China have argued that they should be allowed to use fossil fuels to develop. But their doing this safely would require the Global North to decarbonize first. Yet the United States has remained committed not just to sustaining, but to expanding fossil-fuel production, while blaming the world's lack of climate progress on India's and China's actions.

Even under Democratic administrations, the United States has consistently worked to undermine the UNFCCC regime, beginning almost as soon as it was established. The first international agreement to arise out of the UNFCCC was the 1997 Kyoto Protocol. Kyoto was a legally binding treaty constructed on the UNFCCC's first principle of global equity. It required industrialized, high-income nations, which at the time emitted fully two-thirds of annual greenhouse gas pollution, to reduce emissions to an average of 5 percent below 1990 levels by the end of 2012, while exempting developing and low-income nations from cutting emissions during this period so they could prioritize economic development. Although the United States signed the Kyoto Protocol in 1998, President Bill Clinton did not submit it to the Senate for ratification, knowing that the Protocol would be rejected. The Senate was in Republican hands, yes, but the Democrats were fully on board with Republican climate obstructionism, having the year before helped their colleagues unanimously pass a resolution (now called the Byrd-Hagel Resolution after the bipartisan pair of senators who proposed it) declaring that the United States should not become a signatory to any international climate agreement that would mandate limits on US greenhouse gas pollution unless it also required "new specific scheduled commitments to limit or reduce greenhouse gas emissions for Developing Country Parties within the same compliance period."[12] Although the United States had ratified the UNFCCC in 1992 under the first President Bush, it now flatly rejected the treaty's equity considerations.

In one sense, Byrd-Hagel simply articulated a standard American liberalism, in which responsibilities, like rights, are held to be universal, requiring all nations to reduce emissions equally. But in another sense, Byrd-Hagel was a startling expression of American exceptionalism. As the nation with the largest historical emissions, at the time the largest annual emissions, and the largest economy on earth, the United States bore the greatest responsibility for the climate crisis and enjoyed the greatest capacity to decarbonize. Yet the senators, both Republican and Democratic, refused to acknowledge these historical facts, showing no moral concern that by continuing to take more than its fair share of the carbon budget, the United States might be damning poorer nations to suffer the worst harms of climate change. Doubtless every nation tries to protect its own interests and maximize its geopolitical power. And certainly in the 1990s climate change was thought to be only a minor threat to wealthy countries, and to the United States in particular. As we saw in "Cost," George W. Bush's Press Secretary Ari Fleischer said that the Kyoto Protocol was "not in the economic interests of the United States," largely because William Nordhaus was assuring policymakers that 3°C of warming by 2100 would be economically optimal for America. (Insofar as it used data, Nordhaus' model took its "empirical evidence on the costs of reducing emissions . . . and on the damages from greenhouse warming" from, as Nordhaus put it, "data *for the United States*.")[13] In 1997 US Senators may well have thought that they had no compelling reason to help halt global heating. But by passing a Resolution that promised *never* to ratify a climate treaty that failed to place equivalent obligations on developing nations, the Senate undermined the structure of global climate governance and effectively codified its commitment to its own fossil-fuel powered economy.

This commitment has since been upheld by Republicans and Democrats alike. If George W. Bush formally withdrew from Kyoto in 2001, and Donald Trump pulled out of the Paris Agreement in 2020, Presidents Obama and Biden, while participating in global climate negotiations—and even setting themselves up as the leaders of the international climate regime—have worked behind the scenes to prevent the formation of any legally binding commitment to reduce emissions or phase out fossil fuels, while placing the responsibility for obstructionism squarely on the shoulders of developing countries. When Obama clinched the Democratic nomination in 2008, he promised in his acceptance speech that history would look back on this event as "the moment when the rise of the oceans began to slow and our planet began to heal." But he refused to countenance the idea that developed countries, and the United States in particular, should unilaterally reduce their emissions.

Every year the UNFCCC convenes a Conference of the Parties, where signatory nations that have signed on to the Convention meet to negotiate emissions reductions and other issues central to halting global heating. At the fifteenth Conference of the Parties (COP15), which took place in Copenhagen in 2009, there was hope that with a Democratic president it might be possible to hammer out a new global agreement to reduce emissions. Yet even before the Copenhagen conference began, talks were foundering upon the revelation, leaked to the British press, that the Danish hosts had been discussing with twenty-five of the wealthiest countries a draft text of a non-binding agreement, separate from the UNFCCC working text, to apportion emissions cuts to everyone, with the goal of peaking global emissions by 2020 and reducing them 50 percent by 2050.[14] India, China, Brazil, and South Africa—which under the banner of "BASIC nations" had recently committed to acting as a bloc in Copenhagen—released a statement making it clear that they would not sign this agreement. Apportioning emissions cuts to everyone, they argued, would mean that developed countries would do less than their equitable share and then "the remaining cuts," as South Africa's chief climate negotiator put it, "must be done by developing countries."[15] The rest of the G77, the United Nations coalition of Global South developing nations, shared the BASIC position and were outraged at the secret discussions and insufficiently equitable draft. Once the conference actually started, it became clear that the Danes had lost credibility, as they struggled to guide negotiations over key issues. When Global North heads of state, including Obama, arrived in Denmark to sign the agreement they had assumed would be forthcoming, no agreement was close to being finalized.

The morning after Obama's arrival, the conference began to fall apart. Abandoning standard procedure and any pretense to inclusivity and transparency, twenty-five Global North and BASIC world leaders retreated to a side room and tried to hammer out some sort of agreement that would allow them to return to their home countries and say that they had acted on climate. In a leaked audio recording of that meeting, the Europeans, German Chancellor Angela Merkel in particular, can be heard pushing for an agreement that included legally binding universal targets for emissions reductions in 2020 and 2050, but the BASIC nations held firm in their position that such targets would be inequitable and therefore unacceptable. The Chinese Deputy Foreign Minister, He Yafei, can be heard explaining that he was "trying to go into the arguments and debate about historical responsibility," because "in the past two hundred years of industrialization, developed countries contributed more than eighty percent of emissions" and "whoever created this problem" should be held "responsible for the catastrophe we are facing." The audio suggests that at this point he was shouted down—he plaintively interjects "wait, hear me out!"—and soon after asks for a "suspension" for a "few minutes for consultation," which effectively ends the meeting.[16]

At this point, it seems, Obama was no longer in the room. The American president had remained mostly silent during these negotiations, interjecting only to say that any climate finance from the Global North would be tied to emissions commitments by developing nations—and that he would not stay in Copenhagen even one more day, even if the negotiations remained unresolved, because, as he put it, "all of us obviously have extraordinarily important other business to attend to."[17] But perhaps Obama had simply not wanted to engage with Yafei, who was a relatively low-level bureaucrat. According to a Danish account of COP15, Obama's staff had been trying to set up a meeting between Obama and Wen Jiabao, China's prime minister, who was attending the conference but not this side meeting. At some point during this meeting, Obama found out where Jiabao was and simply walked out, proclaiming as he left, "I want to see Wen."[18] Heading directly to the room where Jiabao was reported to be, Obama barged in and found Jiabao meeting with the climate envoys of China,

India, Brazil, and South Africa.[19] No-one present at that meeting has spoken publicly about what was said there, but after a little over an hour Obama emerged and, without consulting the Danes or anyone else, called a hasty press conference where he announced that COP15 had produced an agreement representing "a meaningful and unprecedented breakthrough."[20] Then he promptly left the conference and flew home, claiming that "weather constraints in Washington" required him to return to the United States immediately.[21]

It's important to understand that what Obama called "an agreement" was no agreement at all. UNFCCC rules mandate that any legitimate agreement must be adopted by all Parties to the Convention. The so-called "Copenhagen Accord," the text that somehow had been negotiated and written up in just over an hour, was never formally adopted by all the Parties at COP15. After Obama's departure, the Danes first reassembled the small group of Global North leaders, who discussed the text of the Accord for some hours but made no changes, and then called what became a nightmarish 3 a.m. plenary session in which developing countries furiously refused to adopt what they called an "illegal" document that had been negotiated in "secret."[22] The European Union (EU) also opposed the text Obama had negotiated, saying publicly that "it will not solve the climate threat."[23] After hours of acrimony and chaos, the Parties to the Convention agreed simply to "take note" of the Accord and to report any national emissions targets to the UNFCCC on a voluntary basis.

The Copenhagen Accord undermined the UNFCCC climate regime in three interrelated ways. First, it wiped away the staggered timeline for international climate action. Of course, the Accord did "emphasize" signatories' "strong political will to urgently combat climate change in accordance with the principle of common but differentiated responsibilities and respective capabilities," and it gave developed nations the option to submit 2020 targets to the UNFCCC Secretariat while allowing developing nations simply to "implement mitigation actions."[24] But these genuflections to equity were purely rhetorical. The Copenhagen Accord established a precedent that in effect dismantled the framework requiring countries responsible for climate change to decarbonize first while giving poorer nations carbon space in which to develop. Second, the Copenhagen Accord substituted a loose "pledge and review" process for the Kyoto Protocol's legally binding emissions targets, thereby establishing voluntary commitments as the standard for participation in the climate regime. Finally, the Accord diluted the international community's capacity to measure individual countries' progress toward meeting even their voluntary pledges. Instead of setting out emissions-reductions targets, or designating limits on the atmospheric concentrations of carbon dioxide, the Accord made halting warming at 2°C the ultimate goal of international climate action. This move allowed ongoing scientific uncertainty over the relationship between emissions, concentrations, and temperatures to cloud any clear view of countries' performance in upholding their commitments.

It appears that India, China, Brazil, and South Africa believed, or were persuaded by Obama to believe, that the Copenhagen Accord was in their best interests. Perhaps they had such confidence in their future development, and such faith in the protective power of growth, that they decided they would ultimately benefit, and gain international leverage along the way, by helping the United States—and by extension all rich countries—avoid binding emissions targets. Avoiding binding targets is clearly what Obama thought was in America's interests. (The Europeans had been arguing *for* binding targets just that morning.) Certainly the Byrd-Hagel Resolution had made it clear that even a Democratic Senate would not ratify an international climate treaty with such targets unless all Parties, no matter their differential responsibilities or respective capacities, were bound by them.

Regrettably, most of the American news media failed to analyze the outcome of COP15 in these terms, instead representing Obama as the climate savior who rode in at the last minute and corralled India and China into a compromise.[25] But this narrative was manufactured by the Obama administration itself. In yet another security breach related to the Copenhagen conference, a memorandum on the Obama team's climate-communications strategy was left on a European hotel computer and leaked to the press; the memorandum's first directive said that spokespeople should "reinforce the perception that the United States is constructively engaged in UN negotiations in an effort to produce a global regime to combat climate change," including support for a "legally binding treaty."[26] But the United States never supported a legally binding treaty. Even during the negotiations over the 2015 Paris Agreement, the United States so adamantly opposed a treaty with legally binding targets or enforceable compliance measures that one European diplomat reportedly warned that "if we insist on legally binding, the deal will not be global because we will lose the US."[27]

Obama was still positioning himself as a climate leader going into COP21, the 2015 Conference of the Parties in Paris. He had just announced the final version of his Clean Power Plan, the first ever limits on pollution from US power plants, so his posture as such had just been burnished with some credibility. Yet behind the scenes at the conference the United States fought the use of international law to compel nations to phase out fossil fuels. At one point in Paris the American negotiators even argued against including any temperature targets at all in the Agreement, proposing instead the absurdly hazy formulation of "decarbonization this century."[28] In the end their efforts centered on the difference between two tiny words: "shall" and "should." In treaties, "shall" is understood to establish an obligation to perform the action outlined in an article; "should" has a softer, merely suggestive force. On the final afternoon of the conference, in the plenary hall itself, the United States discovered that the word "should" had been replaced by the word "shall" in the final text distributed by the French, so that the Agreement now said that "developed country Parties *shall* continue taking the lead by undertaking economy-wide absolute emission reduction targets." (The origins of this word change remain a mystery.) The US delegation immediately approached Laurent Fabius, the COP21 President, and adamantly objected, threatening to torpedo the Agreement just as it was coming into harbor.[29] While hundreds of delegates waited in the grand auditorium, the plenary session was delayed for a full ninety minutes while the French and the Americans argued. In the end, backing down and allowing the United States to save face, the French issued a statement that the substitution of the word *shall* for *should* was a "typo" that would be corrected. As one observer reported: "a collective gasp went through the room when the Secretariat read the change, but no one raised an objection," so Fabius gaveled the Agreement through by acclamation and the conference came to an end.[30]

To be sure, the United States was not the only country that wanted to prevent the Paris Agreement from establishing legally binding requirements to reduce emissions. India and China also opposed such targets. And India and China supported the American position that the Paris Agreement's compliance mechanism should be "non-punitive," relying only on "transparency," a kind of name and shame system that would supposedly pressure countries into keeping their climate commitments. (It's hard to imagine a nation as committed to exceptionalism, as fundamentally shameless as the United States, being swayed by such a system, but so be it.) The point is not that the United States was the only nation resisting the phase-out of fossil energy; it's that the American negotiating position was actually allied with that of India and China, even though the United States could not claim that it was still developing or that it lacked the capacity to lead the world into a decarbonized future.

And after Paris, the United States doubled down on fossil fuels. The very week after COP21 ended, Obama signed a bill lifting a seventy-year ban on crude oil exports, nominally in exchange for a five-year extension on tax credits for solar energy development. Within five years crude oil production in Texas' Permian Basin increased by 135 percent, and America became the biggest producer of oil in the world.[31] And lest we think that Obama was strong-armed by the Republicans into enabling more fossil-energy production, or that he was somehow ashamed of his actions, we need only look to his own public statements on the issue. In a 2018 speech at Rice University, Obama bragged that he was actually a great friend to the oil and gas industry: "You wouldn't always know it," he said rather petulantly, "but [oil production] went up every year I was president. That whole, 'suddenly America's, like, the biggest oil producer and the biggest gas [producer]'—that was me, people . . . Say thank you!"[32]

Even before he won the 2020 election, President Biden was more committed to climate action than was his Democratic predecessor. Biden campaigned on being a "Climate President," and after taking office in 2021, his first act was to rejoin the Paris Agreement. While managing the second and third years of the Covid-19 pandemic, Biden focused his first legislative efforts on the development and deployment of clean-energy technologies, ultimately signing the Inflation Reduction Act, the largest federal climate package to date. He also immediately positioned himself on the geopolitical scene as the world's climate leader, hosting a virtual climate summit a few months after his inauguration where forty heads of state—including Indian Prime Minister Narendra Modi, Chinese President Xi Jinping, and Russian President Vladimir Putin—made speeches reaffirming their commitments to the general goals of the Paris Agreement. And he sent his new climate envoy, Senator John Kerry, to travel tirelessly abroad to liaise with his global counterparts.

Yet at the same time Biden gave his support to the fossil-energy industry. In at least one year of his presidency, the United States approved more oil and gas expansion than did any other country, including Saudi Arabia.[33] Officials in his administration sometimes suggested that new US extraction projects, such as ConocoPhillips' Willow project in Alaska, or new fossil-fuel infrastructure, such as the Mountain Valley Pipeline, were political compromises that enabled him to pursue his climate agenda, but Biden also facilitated new fossil-energy projects that had nothing to do with domestic legislation. Before pausing such approvals in early 2024, at the onset of his Presidential re-election campaign—in order to "heed the calls of young people" asking him to do more on climate, as the White

House's announcement put it—Biden's Department of Fossil Energy approved nine massive liquid natural gas export terminals, ramping up permitting especially after Russia's 2022 invasion of Ukraine.[34] These projects were given the green light so that, after they were built, years down the road, the United States could supply Europe and Asia with the methane gas they would no longer buy from Putin, amply demonstrating that Biden's long-term energy plans included expanded fossil-energy exports. Indeed, by the end of 2023, before these projects were even built, the United States had already become the largest methane exporter in the world.[35] The emissions from exports are not added to the total of US emissions, because under UNFCCC rules they are associated with the country of combustion, not production. But one analyst has noted that if they were included in the carbon footprint of the United States, then US emissions in 2030 would be roughly the same as they were in 2005 near their peak.[36] And in any event all carbon-dioxide emissions cross national borders and mix into the atmosphere, raising temperatures everywhere.

More broadly, Biden supported a regulatory environment where American capital could easily flow into global fossil-fuel development. Although Saudi Aramco spends more on upstream investments than any other single fossil-fuel company, the US majors—ExxonMobil, Chevron, and Shell among others—together invest the most money into new extraction overall.[37] And under Biden the biggest financiers of these projects have been American banks. In 2021, for example, the world's banks lent fossil-energy companies $742 billion to build new fossil-energy infrastructure; the four banks that lent the biggest sums were the American lenders JPMorgan Chase, Citi, Wells Fargo, and Bank of America.[38] US banks are even funding new coal-fired power plants. Research by the German non-governmental organization Urgewald finds that US investors collectively account for around 58 percent of institutional investments in the global coal industry, with equity and bond holdings of around $602 billion. The two largest institutional investors into coal are the US mutual fund Vanguard, with holdings of almost $101 billion, and BlackRock, with holdings of $109 billion.[39] The Biden administration has supported these coal investments. At the G7 meeting in 2023, the United States managed to block a proposal by the United Kingdom to set a 2030 deadline for phasing out unabated domestic coal power generation.[40]

The United States also supported coal development at COP26 in Glasgow, Biden's first Conference of the Parties. There the United States allied itself with, yes, India and China to block a proposal—advanced by the EU, many of the G77, and small island nations—for the Conference's Cover decision to call on all nations to accelerate the phasing-out of coal. (The United States is the third largest consumer of coal in the world, after China and India. Per capita, the United States burns three times as much coal as India.)[41] In a closed-door meeting, the United States, China, and India forced the EU to drop its proposal, and the conference's final decision ended up calling only for "accelerating efforts towards the phasedown of unabated coal power."[42] In this statement coal is simply phased *down* rather than *out*—and what is phased down is not coal, but "*unabated* coal." This adjective "unabated" leaves open the possibility that countries could claim without verification to be "abating" or offsetting their coal emissions. Of course, a COP Decision that included the word "coal," abated or not, did make some small progress. Never before had any fossil fuel even been mentioned in a UNFCCC document. But in light of planetary realities, developing and European nations called the weaker term "phasedown" "a profound disappointment" and a "bitter pill."[43]

Despite the role of the United States in its manufacture, this bitter pill was largely blamed on India—partially because in the final plenary, India's environment minister, Bhupender Yadav, was the one to present the change to the whole conference. Attempting to excuse his country's support for coal by appealing to equity, Yadav plaintively asked his fellow delegates: "How can anyone expect that developing countries can make promises about phasing out coal and fossil fuel subsidies when developing counties have still to deal with their development agendas and poverty eradication?"[44] Yet it is hard to see why Yadav spoke up for the whole group. India had been showered with accolades at the beginning of the conference for making its first ever net-zero pledge, so perhaps Yadav was just attempting to retain some good will. His attempt at self-justification backfired, however, suggesting to many that India, along with China, should be the exclusive targets for criticism. Even the President of the COP, the United Kingdom's Alok Sharma, said not one word about the United States and instead fumed publicly that "China and India—they will have to explain to climate-vulnerable countries why they did what they did."[45] Few seemed to notice that Biden's climate envoy John Kerry spoke up in support of the weaker language on coal, telling a questioning journalist, with some pique, that "you have to phase down coal before you can—quote—end coal."[46]

Despite Kerry's apparent incrementalism, something seemed to shift between COP26 and COP27. At the following year's climate conference in Egypt, the United States took the unprecedented step of supporting language that targeted not only coal, but also oil and methane gas. These latter two are the dominant fossil fuels in the United States. India, still concerned with global equity, proposed that the conference's Cover Decision call for accelerating the phasedown of not just coal, but all fossil fuels. The United Kingdom, the EU, most of the G77, and all small island nations supported this proposal—as did the United States, to everyone's complete surprise. Certainly, US support was deeply qualified. Kerry emphasized in an interview that the United States had supported only phasing down, not phasing out fossil fuels, and on an uncertain time horizon ("phase down, unabated, over time—the time is a question," were his exact words).[47] And his position may well have been a publicity stunt—the United States surely knew that Saudi Arabia and Russia would strike oil and gas from the final text, as they inevitably did. But still. Any US support for an explicit reduction in oil and methane production was a real first.

At COP27 the United States took another unprecedented step, announcing its support for a "loss and damage" fund to help compensate the most vulnerable nations for the devastations of rising sea levels and extreme weather. Never before had the United States even implicitly acknowledged that its historical emissions were damaging countries in the Global South. The order to change the US position on Loss and Damage reportedly came directly from President Biden through his climate advisor John Podesta. Biden also sought to increase the amount of climate finance that the United States delivers to developing countries. At a 2021 speech to the United Nations General Assembly, Biden pledged that by 2024 the United States would provide developing countries with $11.4 billion annually for climate mitigation. This number represented roughly half the climate finance directed toward the Global South from all sources in 2020.[48] Yet Biden's $11.4 billion was paltry and belated. At the 2009 Copenhagen conference, and then again in Cancun the following year, developed nations pledged to mobilize $100 billion in public and private climate finance annually by 2020.[49] That $100 billion has never materialized.

The reaction to Biden's $11.4 billion pledge illustrates how challenging it can be to mobilize capital for development that leapfrogs fossil fuels. A Democratic Congress voted down Biden's promise in 2023, allocating a mere $1 billion for climate finance in its $1.7 trillion budget. Representative Sean Casten, an Illinois Democrat and a self-avowed climate hawk, spoke for some of his colleagues when he worried that American voters would feel shortchanged by a bigger foreign-aid package: "the scenario where we have Americans who are taxpayers and voters telling us in Congress, 'I need your help,' and we choose to send money to people in Bangladesh or Pakistan or somewhere else, who may need it much more . . . it was one of the hardest things for me at this COP."[50]

But what Democrats like Casten—and Biden, for that matter—should communicate to American taxpayers and voters is that climate finance is not merely philanthropic. Nor is it, as Donald Trump once claimed, "yet another scheme to redistribute wealth out of the United States."[51] Rather, it is an essential element of global climate mitigation and, as such, a key pillar of US self-interest. Global heating will increasingly harm Americans' health, security, and prosperity until the entire world stops adding greenhouse gases to the atmosphere. Developing countries need the resources and the capacity to build power systems centered on wind, solar, nuclear, green, hydrogen, hydropower, and other sources of climate-safe energy. Even India mostly lacks the broad capital base for large initial-outlay investments into green infrastructure and industry. That is why, in 1992, the UNFCCC codified the expectation that wealthy nations would "take the lead" not only by decarbonizing first, but also by providing financial assistance and technology transfer to poorer countries to ensure they develop along the lowest carbon pathway possible.

And what was true in 1992 remains true today. The rapid expansion of the fossil-fuel economy since 1980, which has added more carbon dioxide to the atmosphere than had been added in all previous human history, has not created anywhere near enough wealth in the Global South to obviate the need for capital flows from the Global North. The net-zero pledge that India, for example, presented at COP26 included a request for $1 trillion in financing by 2030.[52] This may seem like a lot of money, but it's actually a bargain—approximately 1 percent of one year of Global GDP, spread out over nearly a decade. This is a tiny price to pay to help the world's third largest emitter reach net zero while bringing people out of poverty. For if countries in the Global South fully industrialize using fossil energy and global temperatures keep rising, the consequences will be catastrophic for all.

China Wants to Be the Solution

Sometimes it can be hard to see China as a nation with any climate ambition whatsoever. From many angles China appears to be a coal state whose authoritarian leader Xi Jinping is more than willing to burn the planet to ash if it consolidates his own power at home and abroad. Xi has been very clear that his overarching goal is to make China the nation "that leads the world in terms of composite national strength and international influence by the middle of the century."[53] Xi announced this goal in his speech to the twentieth Chinese Communist Party (CCP) Congress, where he also won an unprecedented third term as China's President and performed the vanquishing of competing CCP factions by having former President Hu Jintao forcibly escorted out of the Congress's closing ceremony.

Xi has cemented alliances with other autocrats in order to build geopolitical coalitions that might counter American power—and in order to supplement China's domestic coal stores with new supplies of oil and methane gas. In December 2002 Xi met with Saudi Arabia's Mohammed bin Salman to finalize what China calls a "comprehensive strategic partnership," in which China will provide the Kingdom with cloud computing and high-speed internet infrastructure through the state-run technology company Huawei (which the United States had sealed off from its markets for telecom equipment) in return for favorable deals on fossil fuels, such as the 480,000 barrels per day of crude oil that Saudi Aramco promised in 2023 to deliver for twenty years, paid for, notably, in yuan rather than dollars.[54] Looking north as well as west, Xi signed a "comprehensive strategic partnership of coordination for a new era" with Vladimir Putin, and announced that the "friendship" between China and Russia has "no limits." Xi then visited Moscow despite Putin's having been charged with war crimes by the International Criminal Court just days before. Defying Western sanctions on Russian fossil fuels, China has signed multiple deals with Gazprom, the Russian state gas company, buying billions more cubic meters of their liquid methane gas, while also importing record amounts of Russian oil, helping Putin find new markets for the fossil energy he no longer sells to Europe.[55] In these moves China is clearly embracing anti-democratic geopolitics and long-term fossil-energy commitments.

Yet at the same time, China seems to be preparing for a future without fossil energy. It has by all accounts cornered the global market for clean-energy technologies. Very early on China seems to have understood the economic and geopolitical potential of phasing out fossil fuels, and it quietly implemented an industrial strategy that would make the world largely dependent on China when it was finally ready for climate action. In October 2010, less than a year after COP15 in Copenhagen, China's State Council announced that the country would accelerate the development of seven "Strategic Emerging Industries," as they were called, including "energy-saving and environment protection," "new energy" (China's term for what Americans call "clean energy"), "new materials," and "new-energy cars."[56] To nurture these industries, Beijing provided fledgling manufacturers with generous subsidies, tying this financial support to local content rules to ensure domestic growth.[57] China also skillfully negotiated the international trade tensions that arose from its protectionist industrial strategy, once escaping sanctions by the World Trade Organization in part by threatening to impose a tariff on European wine.[58] By the 2020s, China had become the producer of three-quarters of all solar panels and well over half of all wind turbines.[59] China also managed to dominate the production of clean-energy materials. By 2021 it manufactured over three-quarters of the world's polysilicon and 97 percent of the semiconductor wafers used to make solar power cells.[60] It also won control over roughly 60 percent of the market for lithium and other metals used in electric-car batteries and renewable-energy storage.[61] By 2022 it was producing around half of all electric vehicles sold across the world.[62] Not only manufacturing climate technologies for export, China is also deploying those technologies on a staggering scale at home. By the end of 2021 China's installed solar and wind energy capacity accounted for around 40 percent of the global total.[63] The total amount of power that China generated from wind and solar in 2022 was greater than the total power generation of France and Germany together.[64] In 2023, China installed more solar panels in one year than had the United States over its entire history to date.[65] This astonishing rate of deployment shows no signs of slowing.

The acceleration of new-energy deployment in China followed directly on the heels of Xi's historic announcement at the September 2020 United Nations General Assembly that China would peak its emissions before 2030 and achieve the neutrality of not just carbon dioxide, but all greenhouse gases before 2060. As soon as Xi announced these "dual carbon goals," as the Chinese call them, not only did China begin to roll out renewables at unprecedented speed, but the CCP started to integrate climate policy into every aspect of Chinese governance. In January 2021 the Chinese Ministry of Ecology and Environment released "Guiding Opinions on Integrating and Strengthening Efforts in Climate Action and Ecological and Environmental Protection," meant to provide advice to the CCP as it developed new climate policy.[66] In March, the People's Bank of China announced its aim to "comprehensively factor climate change in its policy framework," including monetary policy, risk disclosure, and foreign-exchange reserves.[67] In June, Xi created a new "Leading Small Group" (or, depending on the translation, "National Leading Group") on Climate Change, Energy-Saving, and Emissions Reduction.[68] (A Leading Small Group is a high-level CCP committee typically used to administer broad and complicated issues that cut across the jurisdictions of multiple ministries.)[69] Notably Xi housed the climate Leading Small Group under the aegis of the National Development and Reform Commission (NDRC), the agency that manages China's national economy, which suggests that Xi was integrating climate policy into macroeconomic policy planning. Xi also started to consider climate in China's geopolitical moves, announcing at the 2021 United Nations General Assembly that China would no longer build coal plants abroad. This "no new coal overseas" policy had an immediate impact on the construction of new coal power: in the year after Xi's announcement, around fifteen China-backed overseas coal projects were shelved or canceled due to withdrawals of Chinese firms.[70]

All this led to the publication, in October 2021, of China's overarching policy framework for its carbon peaking and decarbonization. Called the "1+N" framework, it lays out the goals of Chinese climate policy and the structure of Chinese climate governance. The "1" document—the "Working Guidance for Carbon Dioxide Peaking and Carbon Neutrality in Full and Faithful Implementation of the

New Development Philosophy" (the Guidance)—describes China's overarching vision for decarbonization. The first "N" document—the "Action Plan for Carbon Dioxide Peaking before 2030" (the Action Plan)—represents the first iterative plan for China's transition to carbon neutrality. As if they've been afforded the highest importance, the "1+N" policies are now prominently displayed on the NDRC website next to the Fourteenth Five-Year Plan, which outlines China's strategy for economic development between 2021 and 2025. As the Action Plan proclaims, "the goal to peak carbon dioxide emissions permeates the whole process and every aspect of [China's] economic and social development."[71] The "1+N" documents take a whole-of-government approach to creating a carbon-neutral China, including climate policies to be implemented by all national and provincial ministries, agencies, and institutions. The documents are long and comprehensive, offering a detailed blueprint for a full systems change to a net-zero economy. What follows is an overview of this blueprint—or, if you prefer, of the vast architecture of China's new climate policy.

The high-level vision, the Guidance, describes how China wants to transform its energy, industry, transportation, agriculture, and built environment. It declares that China "will" substitute clean energy for fossil fuels: "We will carry out initiatives to substitute renewable energy for fossil fuels, vigorously develop wind, solar, biomass, marine, and geothermal energy sources among others, and continuously increase the share of non-fossil energy in total energy consumption . . . prioritizing local development and use of wind and solar power."[72] To tackle emissions from industry, the Guidance proclaims that China will actively phase out polluting industries, phase in new low- and zero-carbon industries, and develop new zero-carbon economic sectors. It insists that "capacity substitutions must be strictly implemented" in the production of high-emissions materials like steel and cement. Turning to emissions from agriculture, the Guidance directs the bureaucracy and private industry to promote green agricultural development and "stabilize the carbon sequestration function of existing forests, grasslands, wetlands, seas, soils, permafrost, and karst areas." It affirms China's determination to build a low-carbon transportation system centered on electric vehicles and electrified public transportation. It requires that municipalities and city governments apply "green and low-carbon requirements" to "every link of urban and rural planning," including by retrofitting existing buildings. And it states that the central government will initiate "major projects for protecting and restoring ecosystems."

Next the Guidance lays out the blueprint for China's climate governance. It announces that China will write legislation for the benefit of green development and will "remove the contents in existing laws and regulations that are incompatible with the task of carbon dioxide peaking and carbon neutrality." It notes that China will implement "preferential tax policies for environmental protection, energy and water conservation, new energy, and clean-energy vehicles and vessels," while encouraging "sound pricing mechanisms" for promoting "the large-scale development of renewable energy." In order for these tax policies to function, China will, the Guidance emphasizes, "establish a system to control the total volume of CO_2 emissions" by having officials "build [China's] statistical and accounting capacity for CO_2 emissions and enhance the quality of measurement through the use of information technology."

The Guidance also mobilizes capital, directing China's financial institutions to "build investment and financing systems tailored to the goals of carbon dioxide peaking and carbon neutrality." Integrating climate into China's trade policies, the Guidance declares that the NDRC will create a "green trade system," restrict exports of "energy-intensive and high-emission products," and make "green development" a "defining feature" of the Belt and Road initiative. The Guidance also prioritizes innovation, proposing an "open competition mechanism to select the best candidates to lead research on low-carbon, zero-carbon and carbon-negative technologies and on new materials, technologies, and equipment for energy storage," including "controlled nuclear fusion."

To seed this innovation and promote cultural transformation, the Guidance proclaims that China will integrate carbon neutrality into public education. China will, it says, "incorporate green and low-carbon development into [its] national education system" to "build societal consensus." It will especially "encourage universities and colleges to establish disciplines and majors relevant to peaking carbon dioxide emissions and achiev[ing] carbon neutrality." Further, this education will "advocate simple, moderate, green, and low-carbon living patterns." (In a similar vein, the Action Plan announces that in the five-year period that ends in 2025, China will attempt to "curb luxury, waste, and unnecessary consumption," intensifying "initiatives to promote eco-friendly living patterns" and selecting and publicizing "a group of role models" who can inspire others to adopt a low-carbon lifestyle.)

Finally—and, in Chinese climate politics, above all—the Guidance signals that China's climate and ecological goals must be fully embraced by all members of the Chinese Communist Party. "Carbon dioxide peaking and carbon neutrality will be an important part of the officials' education and training system," the Guidance proclaims, so that "officials at all levels" can "effectively promote green and low-carbon development." Indeed, the Guidance exhorts CCP leaders in provinces, municipalities, and territories to "resolutely shoulder their responsibilities," "set clear goals and tasks," and "formulate implementation measures," building "targets for CO2 peaking and carbon neutrality" and making them "more binding."

The Guidance clarifies that the work of achieving carbon neutrality in 2060 must begin immediately—and the Action Plan lays out the details: clean energy must be 20 percent of overall energy consumption by 2025, 25 percent by 2030, and over 80 percent by 2060. (For comparison: in 2021 renewables generated close to 22 percent of total energy consumption in the EU and 20 percent in the United States.)[73] And the Guidance establishes that these and indeed all China's climate targets are subject to assessment and verification: it commands that "all local authorities and relevant departments" should "annually submit implementation reports to the CPC Central Committee and the State Council," and reiterates that these reports will be verified by "the Central Inspections on Environmental Protection." It promises that the "outstanding regions, organizations, and individuals" who meet their climate targets will be "duly rewarded," and it warns that the "regions and departments that fail to accomplish their goals" will be "criticized by means of circular, called in for talks, and held accountable with laws and regulations" enforced by "all local authorities."

This summary has listed only a fraction of the policies, targets, and goals set out by the Guidance and the Action Plan. Taken as a whole, the "1+N" documents present a complete, and powerfully stated, climate agenda more ambitious than any advanced by a Global North nation. Yet of course China's climate policies sit uncomfortably next to some of China's energy investments. Even as the "1+N" documents were being written, China was approving domestic coal-fired power plants at a feverish clip. China's domestic coal development was responsible for over half the coal development across the globe in 2021.[74] In 2022 China issued permits that together will generate six times as much coal as the rest of the world combined.[75] In the first three months of 2023 it permitted more new coal power plants than in the whole of 2021.[76] Rather than pulling away from fossil energy, in these years China seemed to be embracing it with ever more enthusiasm.

It is important to know, however, that China began building new coal capacity because it did not have enough energy in the summers of 2021 and 2022 to respond to devastating, climate-fueled heat waves. Before these heat waves, Chinese coal development had been on the decline.[77] But the back-to-back years of extreme weather drove China's hydropower reserves to some of their lowest historical levels while simultaneously spiking nationwide demand for air conditioning. In 2021 two out of three Chinese provinces suffered weeks-long blackouts.[78] Not only did factories shut down, but hundreds of millions of households lost power. Surely some people died. The CCP had to act, but China did not yet have enough transmission in place to support a full build-out of renewables. So the CCP decided to use their domestic coal stores to establish more power capacity in preparation for those weeks of peak demand (even if those coal plants would not run near full capacity, or at all, during other times of the year).[79] It may have been wildly inefficient, but this was their decision.

China's increasing its coal capacity does not mean, however, that the "1+N" documents are nothing more than greenwashing. The "1+N" documents speak quite frankly about what China is actually doing. The Action Plan makes it very clear that until 2030 the "bottom line" for China is "national energy security and economic development." The reason China has "dual" carbon goals in the first place is that it needs, it says, more time to establish modernized infrastructure and safeguard "China's energy security, food security, and the security of industrial and supply chains" before it can "realize the gradual replacement of new energy" while keeping "ordinary citizens living and working as normal."[80] It might stretch credulity to imagine that China might need to modernize any further, but despite its rapid development over the past decades, per capita Chinese income remains below roughly $15,000 per year, which by some measures means China remains part of the underdeveloped Global South.[81] And given the comprehensiveness of the "1+N" policy framework, it seems reasonable to assume that the Chinese Communist Party assessed the trade-offs of building more coal now and decarbonizing by 2060. European analysts have found that, so long as the rest of the world upholds their own climate pledges, the "1+N" policies could halt warming at 2°C.[82] China has apparently decided that 2°C of heating is an acceptable cost of temporarily continuing to grow per capita GDP with fossil-fuel-powered industry. Meanwhile China seems to be preparing to pivot toward carbon neutrality. In 2022, China invested eleven times as much money in renewable power as it did in coal.[83] And in 2023 it announced for the first time ever that non-fossil power sources accounted for over 50 percent of China's total installed electricity capacity, a greater amount than the 40 percent of such installed capacity in the United States that year.[84]

Of course, even if the CCP throws its full weight behind decarbonizing their economy by 2060, implementing the "1+N" policies will be extremely challenging. Any complicated legislation is subject to the fog of enactment, and this fog is all the thicker in China, where policies issued by the CCP, and rubber-stamped by the State Council, rely on the nation's vast bureaucracy for their implementation.[85] China's provincial and territorial governments are part of a complex hierarchy of supervision and subordination, with regional administrations both controlling their own budgets and running key state-owned enterprises, many of which are in high-carbon industries like cement.[86] In the past there have been chronic problems enforcing the central government's environmental initiatives. As the CEO of London's China Dialogue Trust Sam Geall points out: "At local levels of government, contradictory laws, collusion between officials and polluters, misaligned political evaluation metrics for officials, and restricted scope for citizen oversight have all thwarted environmental initiatives. At the elite level, vested interests, interagency rivalries and an overriding focus on high growth rates . . . have worked against green policies."[87] It remains to be seen if the new "1+N" climate governance will ultimately succeed in managing the interests of party factions and officials throughout the bureaucracy.

The need to accommodate governmental fragmentation and factionalism means that the language of Chinese politics is characterized by a remarkable degree of ambiguity and flexibility.[88] The Johns Hopkins Alfred Chandler Chair of Political Economy Dr. Yuen Yuen Ang has proposed that there are three kinds of national policy directives in China: "grey," which are flexible or ambiguous about what can or cannot be done; "black," which clearly state what will be done; and "red," which clearly state what must not be done. "Together," she says, "this mixture of ambiguous and clear directives forms a system of adaptive policy communication."[89] In other words, the flexibility of Chinese policy language allows different factions and governments in China's fragmented authoritarian hierarchy to adapt the CCP's directives to suit their own particular needs, while giving plausible deniability to the central government when their directives are not strictly followed.

Yet that kind of ambiguity does not, in fact, characterize the language of Chinese *climate* policy, which, throughout the "1+N" documents, speaks in starkly unequivocal terms, saying that China "will" do X or "will have done" X by a certain date. This kind of explicit goal-setting has a special political import in China, where (at least over the past decades) the CCP has derived some legitimacy from its performance—its state capacity to do what it says it is going to do.[90] This performance-based legitimacy is different from procedural legitimacy in the United States, where elected officials may fail to secure their policy priorities, or decline even to attempt to enact their "campaign promises" (which voters generally understand to be advertisements rather than commitments), yet at the same time the system itself remains legitimate because it manifests the people's consent to the democratic process. China does not have this democratic participation, just the supposed hypercompetence of benevolent authorities. Hence the propaganda of Chinese state media constantly emphasizes the honesty of China's leaders and the efficacy of its government. In a commentary about China's climate plans, for instance, Zheng Zeguang, the Chinese ambassador to the United Kingdom, reiterates the party line: "Anyone familiar with China's political system knows that once decisions and goals are set by the CPC central committee and the top leader, they are incorporated into the overall national development program, turned into feasible action plans and delivered faithfully by local governments and competent departments. That is how the country has achieved its development miracle over the seventy-two years since the founding of the People's Republic."[91]

Obviously Xi and the CCP can control public perceptions to some degree because, by controlling not only state but also social media channels, it is relatively easy for them to cover up their policy failures. Yet the UCLA Professor of Law and China scholar Alex Wang has suggested that hiding environmental policy failures is difficult for the CCP when the effects of their failures are visible.[92] Wang points to air pollution in Chinese cities as the kind of environmental problem that the CCP has been forced to address.[93] (After air pollution became a leading source of protests in both rural areas and cities, China lowered the quantity of harmful particulates in its air by as much as 40 percent, reducing smog nearly as much in seven years as the United States did in three decades.) So far, as Wang notes, climate change has been only ambiguously visible, but it will become ever more present as the planet warms. And China's vulnerability to climate change is a serious concern for the Chinese public. A European Investment Bank survey has found that 73 percent of Chinese respondents consider climate change a major threat to their society, and 94 percent believe climate change is already impacting their daily lives.[94] Similarly, the Yale Program on Climate Communication has reported that 94 percent of Chinese respondents feel that climate change is important to them personally. The Yale researchers have also reported that a nearly unanimous 97 percent of Chinese respondents support China's dual carbon goals, and an extraordinary 90 percent believe that these goals "can be achieved."[95] Even if the actual numbers of Chinese people with these feelings and beliefs are smaller than these surveys suggest, it seems clear that majorities expect that the Chinese Communist Party should, and can, decarbonize China—and that Xi will need to address climate change to sustain his power.

How (Not) to Talk about "India and China"

The next time someone claims it doesn't matter whether the United States acts on climate because emissions are rising in "India and China," you can regain control of the conversation by citing China's enormous rate of clean-energy deployment and by describing the scope and ambition of its climate policies. Then you can redirect the discussion by pointing to what the "India and China" tactic is designed to obscure: the United States' obstructionism in international climate negotiations and its ongoing commitment to fossil-fuel development. This destructive behavior on the part of the United States is what the language of climate politics should highlight.

To inspire calls for the United States to change, you can frame China's climate action as a threat to American hegemony. In this frame, Chinese support for clean energy and its comprehensive climate-policy commitments can be represented as elements of Xi's ambition to have China "lead the world in terms of composite national strength and international influence by the middle of the century."[96] As President Biden's National Security Advisor Jake Sullivan said in a 2021 speech at the non-partisan United States Institute of Peace: "China is essentially making the case that the Chinese model is better than the American model. They are pointing to dysfunction and division in the United States and saying . . . their system doesn't work. Our system does."[97] You can mention that Chinese state media repeatedly criticizes America for its climate polarization and paralysis. For example, after the Supreme Court limited the Environmental Protection Agency's capacity to regulate carbon emissions from power plants, a Chinese spokesperson, Liu Pengyu, sniped that "these self-contradictory moves make the world question the US's capability and seriousness."[98] If China achieves its dual carbon goals while the United States remains stymied by its fossil-fuel interests, not only will China continue to dominate the climate technology market, the key sector of any future prosperity, but it will also establish authoritarianism as the form of governance that can best resolve the climate crisis. This would be a tragedy for human freedom that American leaders should do everything to prevent.

If your interlocutor remains skeptical that China will actually implement its "1+N" policy framework, you do not need to change your core message that America must accept its responsibility to lead the world out of the fossil-fuel era. For if China does not decarbonize, the world will have all the more need of American leadership. Taking up that leadership, the United States could compel Chinese cooperation by establishing a legally binding treaty to phase out fossil fuels and an international trading bloc to sanction nations that refuse to become party to it. The treaty would need to exist outside the UNFCCC regime, because under UN rules any climate agreement must be signed by all parties (which gives any one state the power to block any agreement which it deems counter to its interests). A fossil-fuel treaty could include, at first, the United States, the EU, and the majority of the G77—the nations that have been asking the United States to make legally binding emissions commitments since Kyoto. To shift the trajectory of Asian development, the treaty would soon need to include India. Arguably India could be brought on board with a kind of climate Marshall Plan: a treaty-established fund for India's investments in solar farms, flexible power grids, energy-storage facilities, and so on. Right now India faces exorbitant costs of capital—financing to build a solar farm in India is three times as expensive as it is in Europe.[99] An international mechanism that made phasing out fossil energy an immediate economic benefit to India could be a global game changer. And, you should make sure to say, this treaty would benefit the world's wealthiest countries too. All nations, rich and poor, will be increasingly harmed by global heating for as long as it's allowed to continue. The poor are being harmed first and worst, but if the world does not phase out fossil fuels, in the end climate change will come for everybody.

The United States in concert with treaty signatories could also establish an International Hydrocarbon Agency to manage the phaseout of fossil fuels across parties and establish border carbon adjustments to equalize prices based on the carbon content of goods traded with non-party nations.[100] (This is well within the realm of the possible: the EU has already enacted a European Carbon Border Adjustment Mechanism, and a 2023 poll found that 74 percent of registered American voters support this sort of mechanism.)[101] Joining a treaty with a very strict carbon border adjustment might be especially appealing to India, which has already placed customs duties and tariffs on Chinese solar panels in order to nurture its own domestic solar industry. Columbia University's Jason Bordoff and Harvard University's Meghan O'Sullivan together have suggested that, over time, border price adjustments meant to level the playing field could morph into tariffs aimed at pressuring resistant nations to pursue climate policies. "Although the idea of using sanctions to compel speedier decarbonization may seem over the top now," Bordoff and O'Sullivan argue, "in a world in which carbon emitters are increasingly seen as threats to international peace and security, sanctions could become a common tool to force laggards to act."[102]

Of course, it is all very well to speculate about the ways the United States could phase out coal, oil, and gas while creating a trading bloc that could force China to decarbonize, but you might be confronted with the problem that currently in America itself both Democrats and Republicans support fossil fuels. Yet there are clear and crucial differences between the parties that you could easily emphasize. Republicans remain totally committed to blocking the enactment of any policy that might acknowledge how fossil fuels destroy the habitability of our planet—no matter that their climate denial hurts the economic interests of their constituents or the national security interests of the United States. Even Republicans who acknowledge the reality of climate change put fossil-fuel fealty ahead of any other considerations. Utah is one of the US states with the largest reserves of lithium, and its centrist Senator Mitt Romney once tweeted that we must mine "here at home" because "we cannot afford for the Chinese Communist Party to be the global leader in energy production."[103] Yet Romney joined the rest of his party in voting against the 2022 Inflation Reduction Act, even though the law sent money to Utah to help establish the state's lithium mining industry.[104] This is only one example of the way that Republicans govern at the service of fossil energy interests. Unfortunately, there are many more.

In contrast to their counterparts on the right, Democratic lawmakers by and large support the development of clean energy and the decarbonization of industry, even as they maintain their false and dangerous belief that the world can halt global heating no matter whether America keeps producing fossil fuels. Democrats also remain accountable to an electorate whose views on the climate crisis have been relatively less warped by Fox News–style disinformation. And Democrats do change their positions in response to voter pressure. Biden began his presidential campaign in 2019 with a grade of F from the youth-led climate advocacy group Sunrise Movement, yet he transformed his policy priorities in response to activist demands and what seemed like a sudden but deep scientific education on the climate crisis. Although he continued to support increasing exports of oil and gas even after his climate awakening, Biden also managed to shepherd the Inflation Reduction Act through Congress with only a razor-thin majority in the Senate. Had his Senate majority been larger—taking away the de facto veto power of Joe Manchin, the Democratic coal-baron senator from West Virgina—Biden would surely have won even greater climate victories. And his decision to pause permitting approvals of new liquid methane export terminals is enormously revealing. Biden is not an arsonist, as Trump surely would be as President. Biden is wavering between supporting fossil fuels and leading the world into the twenty-first century. For this century not to herald utter climate catastrophe, Biden must be re-elected—and then pushed to treat the climate crisis with the unequivocal urgency it deserves.

Democratic engagement is the most powerful counter to the cynicism and hopelessness that fossil-fuel interests intend to inspire with the "India and China" tactic. And it thus serves as the culminating message that you can use to fight this form of propaganda. When people feel helpless, you can give them something to do! Donating to organizations working to protect the right to vote, registering voters, helping voters get to the polls, even agitating for statehood for Puerto Rico and Washington, DC: these are all forms of climate action. Donating to climate organizations, participating in public climate protests, repeatedly asking your representatives to withdraw their support for fossil fuels: these are all forms of democratic participation. Getting Democrats elected to the Presidency—and to the Senate, especially in purple states such as Georgia, Arizona, Wisconsin, North Carolina, and Florida—and then pressuring elected Democrats to support the phase out of fossil fuels, are actions that directly influence the international politics of climate change. Using electoral politics to staff our governments with people who could in fact represent our will—and then forcing these representatives to break with fossil energy—will renew America as the nation that leads the world into the next phase of human history. We'll know we're on our way when the Senate passes a new Resolution, superseding the Byrd-Hagel Resolution, which acknowledges that climate change is an emergency and commits the full power of the United States to resolving our planetary crisis. Let's work to enable such a resolution to pass—even as a goal of domestic climate politics—using our power as American citizens to shape policy and help create a livable future.

5

Innovation

The history of [carbon capture and storage] has been that of great disappointment.

—Fatih Birol, International Energy Agency
("Carbon Capture Desperately Needs a Reality Check")

High dependence on [carbon capture and storage] and [carbon dioxide removal] is most likely driven by inadequate model representation of real-world constraints on their potential.

—Ploy Achakulwisut, Stockholm Environment Institute, et al.
("Why All Fossil Fuels Must Decline Rapidly to Stay Below 1.5C")

If oil companies are investing in [direct air capture] so they can keep producing oil and keep drilling, that runs counter to what the IPCC is saying we have to do. We have to phase out fossil fuels.

—Angela Anderson, World Resources Institute
("A Startup Battles Big Oil for the $1 Trillion Future of Carbon Cleanup")

The idea that technological innovation will allow the world to halt global heating while America continues to burn fossil fuels is an idea with great appeal to centrist policymakers. During his tenure as a Democratic Senator from West Virgina, Joe Manchin, for instance, would often tell reporters that he supported climate policies that would lead to "innovation, not elimination"—by which he meant the elimination of fossil fuels.[1] Suggesting that innovation can solve the climate crisis without our ever having to eliminate its main cause, Manchin was of course invoking the Solowian faith that new technologies can always substitute for exhausted natural resources and offset the mounting costs of pollution. But Manchin was also signaling his support for a particular set of technologies: carbon capture and storage (CCS) and carbon dioxide removal (CDR). Indeed, in the language of climate politics the word "innovation" often serves as code for these two technologies. Hence when Congress appropriated funds for carbon-removal research and the construction of carbon dioxide pipelines toward the end of the Trump administration, they called their legislation the "Utilizing Significant Emissions with Innovative Technologies" Act—or the USE IT Act. (After the bill was signed, Wyoming Senator John Barrasso, one of its sponsors, praised Trump for advancing "important legislation" to "reduce carbon emissions through innovation.")[2] And at COP26 in 2021, when the European Commission, the United States, and twenty-one other nations announced a collaboration to "catalyze a global CDR industry," they called their initiative "Mission Innovation."[3] Other new technologies and processes will certainly play a role in decarbonizing the global economy, but when policymakers and governments talk about innovation, they usually mean CCS and CDR.

Carbon dioxide removal is indeed a technological innovation that the world will need to develop and scale in order to achieve net-zero emissions. Even after most of the economy has been decarbonized, some essential sectors like agriculture will continue to emit carbon. These unavoidable emissions will need to be removed by a combination of ecological carbon sinks and human technologies so that they do not add to the stock of carbon dioxide in the atmosphere. But fossil-fuel companies and some corporate and academic CDR advocates are claiming that the world will be able to use carbon removal not just to clean up those marginal, unavoidable emissions, but also to keep consuming large amounts of coal, oil, and methane gas. This claim, I will demonstrate, is false and dangerous.

Yet Democratic policymakers have largely embraced this propaganda, taking fossil-fuel companies' word that they have unique expertise that can innovate and deliver carbon capture and carbon removal, which will empower America, this story goes, to continue to produce domestic fossil energy while meeting its Paris Agreement commitments. Oil and gas lobbyists have successfully used this propaganda, I will show, to siphon billions in fossil-fuel subsidies from the federal government and install their representatives at the table of energy-policy decision-making. And, I will suggest, their lobbying has been so well-received because it has helped Democratic lawmakers justify their ambivalent, all-of-the-above energy policies to themselves.

Of course, the Democrats are not completely naive. They know that carbon capture might not actually work. In 2023 President Biden's then climate envoy, John Kerry, said that "we have to put the industry on notice," warning them: "you've got six years, eight years, no more than ten years or so, within which you've got to come up with a means by which you're going to capture," or else "we have to deploy alternative sources of energy."[4] What Kerry didn't mention was that even in 2023 the world did not have "six years, eight years, no more than ten years or so" to spare while governments dithered and procrastinated on deploying solar and wind power. Global warming will soon overshoot the Paris Agreement temperature targets if the world does not immediately and drastically slash emissions. Continuing to increase the world's greenhouse-gas pollution into the 2030s would be disastrous.

Given this immovable reality, we must counter the fossil-fuel propaganda that carbon capture and carbon dioxide removal will allow the continued wide-scale consumption of coal, oil, and methane gas. Instead, we should advance this clear, simple message and repeat it often: the world must phase out fossil fuels. Governments must wind down the fossil-energy industry, and build up a limited system of carbon removal without their influence, in order to have any chance to achieve net zero, halt global heating, and secure an ongoing future.

The Limits of Carbon Capture and Carbon Dioxide Removal

Carbon removal is part of the Earth's carbon cycle. Trees and other plants pull carbon out of the air, and rocks are weathered by wind and rain in ways that end up storing carbon on the ocean floor. Climate scientists sometimes characterize these processes as "nature-based solutions" to climate change. But nature-based forms of carbon removal are not included in the concept of "innovation." That term is reserved for industrial forms of CDR. There are two main types of industrial CDR: "Bioenergy with Carbon Capture and Storage" and "Direct Air Capture." You will often see their acronyms, BECCS and DAC, respectively, in scientific reports or news articles about climate technologies. Sometimes you will also see the acronym "NET," which stands for "negative emissions technology."

Bioenergy with carbon capture and storage is the form of CDR most often included in modeled "mitigation pathways." (These pathways, which we first encountered in "Alarmist," model how the world might decarbonize the global economy in time to halt global heating at a certain temperature. More on them later.) The idea behind BECCS is to use plants to remove carbon dioxide from the atmosphere, cultivating grasses or quick-growing trees on many millions of acres, then harvesting and burning this so-called "bioenergy" for power generation, while capturing the carbon dioxide released by this combustion, and then, finally, sending the captured carbon dioxide through a network of pipelines to reservoirs where it could be stored. By the end of 2023, around 2 million tons of carbon dioxide per year was captured using this method.[5] This amount represents less than 30 minutes of annual global carbon-dioxide emissions.

BECCS has yet to be scaled, obviously. But it also seems like, at any scale, the technology does not work very well. This is because the part of the process that traps carbon dioxide—the "carbon capture and storage" part of "bioenergy with carbon capture and storage"—is not very effective. Carbon capture and storage, also known as "CCS," is an industrial process in a machine that gets retrofitted onto a refinery or power plant to capture carbon dioxide at the source, preventing the CO_2 from entering the atmosphere in the first place. If bioenergy were to be burned for carbon dioxide removal, its emissions would be captured by such a machine at the site of combustion. CCS has been tried for decades at various sites around the world, but it has consistently failed to capture anything close to 100 percent of emissions. (For BECCS to actually remove carbon from the atmosphere, 100 percent of the emissions from the bioenergy would need to be captured as it's burned, or else the process would just keep emitting CO_2 into the air, if at lower rates.)

The first commercial-scale CCS power plant, the Boundary Dam 3 in Canada, opened in 2014 with the promise that it would capture 90 percent of the CO_2 it produced, but it has run at only 40 percent capacity due to apparently irresolvable technical issues. In 2021, its capture rate was less than 37 percent of its target.[6] (Somehow it sells electricity to ratepayers at twice the standard price, however.)[7] Another CCS project—an Australian liquified methane-gas plant co-owned by Chevron, Shell, and ExxonMobil—opened in 2016 with a goal of capturing 80 percent of its emissions. Bedeviled by technical issues, it didn't start up its capture machinery until 2019, and by 2021 the plant's operators admitted that they had managed to capture only 30 percent of the plant's carbon dioxide.[8]

When the American Petra Nova coal plant opened in Texas in 2017, it was touted by the *Washington Post* as ushering in the era of "clean coal"—which the *New York Times* then called "one of the most promising . . . technologies for fighting global warming."[9] But Petra Nova also failed to meet its capture targets, suffering repeated shut-downs before finally closing in May 2020, having cost $1 billion to build and run—$195 million of which was taxpayer money.[10] Its carbon-capture process used so much energy that an entirely separate methane-gas plant had been built to power it. Thanks to this supplementary fossil-fuel plant, Petra Nova was significantly carbon additive, increasing

America's overall CO_2 emissions by 1.1 million tons between 2017 and 2020.[11] Other CCS projects have also added large amounts of carbon to the atmosphere. While capturing 35 percent of its generated CO_2, the Quest methane reformer in the Alberta oil sands still emitted about 7.5 million tons of greenhouse gases between 2015 and 2019, according to a recent analysis.[12] As Francesco Starace, the CEO of a European utility, once admitted to a reporter: "The fact is, it [CCS] doesn't work, it hasn't worked for us so far . . . and there is a rule of thumb here: If a technology doesn't really pick up in five years—and here we're talking about more than five, we're talking about fifteen, at least—you better drop it."[13]

It is hard to see how bioenergy with carbon capture and storage could actually remove significant amounts of carbon from the atmosphere when the carbon-capture process has yet to achieve its industry target of capturing even 90, let alone 100 percent of emissions. But, as a thought experiment, let's assume that innovation in carbon capture, despite fifteen or more years of failure, manages to iterate a process that captures 100 percent of CO_2 emissions. Might this form of carbon dioxide removal work then? Well, it might—but it might also create significant problems for ecological and human systems. Growing enough bioenergy to capture enough carbon to have any effect on global temperatures would mean planting monocrops on a truly staggering amount of land. According to the National Academies of Science, Engineering, and Medicine, removing 10 billion tons of CO_2 a year from the atmosphere—about a quarter of annual global emissions—would mean planting bioenergy on a land area about one and a half times the size of India.[14] This is nearly half the land area on which the entire planet grows food. Obviously, using arable land to grow bioenergy would spike the price of staple crops—by as much as seven times, in one estimate—causing widespread famine.[15] Yet extracting this land from forest would destroy natural carbon sinks and would itself therefore have deleterious effects on the climate.

Wide-scale BECCS would also do terrible damage to ecosystems. One study warns that growing horizon-to-horizon acres of bioenergy, in addition to whatever crops we would need for food, would lead to "terrestrial acidification, eutrophication, terrestrial ecotoxicity, ionizing radiation and ozone depletion."[16] In other words, it would destroy the chemistry of the soil, choke waterways with fertilizer, create vast dead zones in lakes and oceans, pollute human communities with toxins, expose human bodies to DNA-severing electrons, and worsen the hole in the ozone layer that allows cancer-causing solar radiation to reach the surface of the Earth. And, if that weren't enough, the National Academies points out that "because remaining biodiversity is already threatened by habitat loss exacerbated by climate change, devoting a substantial amount of nonagricultural land to land-hungry NETs would likely cause a substantial increase in extinction." All this is why a landmark 2016 expert survey exploring the feasibility of BECCS concluded that the expectation the technology could remove significant amounts of carbon is "unrealistic."[17]

Scientists' widespread concern about the harms of bioenergy with carbon capture and storage has prompted researchers to look toward a different carbon removal technology: direct air capture, or DAC. This technology remains at a nascent stage of development. As of 2023, only around twenty-seven DAC plants existed worldwide, and together, running for a whole year, they captured less than a minute of annual global carbon dioxide emissions.[18] Many thousands of these plants would need to be built globally, each accompanied by an extensive network of pipelines to transport carbon dioxide to storage sites, for direct air capture to play a meaningful role in a net-zero energy system.[19]

Direct air capture plants are not small. In fact, they are mid-sized industrial facilities surrounded by stacks of open containers, up to seven stories tall or more, holding giant fans that roar loudly as they pull air indoors to be processed. These fans, and the intense industrial heat (about 100°C or 212°F) required for the capture process, also require that large amounts of energy be sited relatively close by, and this energy needs to be virtually zero-carbon. The National Academies has calculated that to run enough DAC to capture 1 million tons of carbon dioxide—around fifteen minutes of annual global emissions—solar panels, as well as a methane-gas plant supplying the thermal energy, would need to be cited on somewhere between 1,355 to 2,450 acres.[20] (The National Academies does not model direct air capture without using fossil fuels, because the production of industrial heat from solar energy is still being innovated.) For a sense of the scale here: an American football field is 1.32 acres. Over one whole football field's worth of land and an additional fossil-fuel power plant would be needed to capture roughly fifteen minutes of annual global emissions.

Direct air capture's thirst for energy also makes this technology enormously expensive. Current estimates for capturing one ton of carbon dioxide range up to $1,000, not including the costs of CO_2 compression, transportation, injection, and sequestration.[21] Presumably these exorbitant costs reflect an early phase of innovation that would be superseded by economies of scale. But there is also the immutable fact that energy costs money, and taking CO_2 out of the air uses more energy than emitting it in the first place. Noting pointedly that "low-cost estimates tend to come from sources closer to industry," the definitive literature review of research into direct air capture finds that thermodynamic considerations alone rule out estimates of costs below $500 per ton of CO_2.[22] If capturing a ton of carbon dioxide will ultimately cost $500, then capturing a quarter of global annual emissions—10 billion tons—every year will cost $5 trillion per year, roughly 20 percent of annual US gross domestic product.

Researchers warn that due to its extreme resource intensity and inordinate economic cost, direct air capture might not be feasible at large scales. A 2020 energy-systems analysis shows that "the energy and materials requirements for [direct air carbon capture] are unrealistic even when the most promising technologies are employed."[23] A 2021 chemical-engineering paper finds, in a typical split between technical and systemic feasibility, that while direct air capture "shows a large removal potential . . . its large heat and power requirements hamper its large-scale adoption."[24] Scientists authoring a comprehensive literature review of CDR research conclude that they "do not advocate presenting negative emissions technologies as another mitigation option," and researchers studying the costs, potentials, and side effects of CDR note that "large-scale deployment of NETs . . . appears unrealistic given the biophysical and economic limits that are suggested by the available, yet still patchy, science today."[25]

The National Academies is more positive about carbon dioxide removal in general, suggesting that the only major challenge to CDR is "high current cost," but at the same time it judges DAC unready for widescale use and estimates that the remaining options (BECCS, reforestation, and regenerative agriculture) will realistically remove significantly less than even 10 billion tons annually, with the additional caveat that logistical challenges imply even these "achievable limits" could be "smaller by a factor of two or more."[26] The European Academies' Science Advisory Council repeats this assessment, bluntly: "having reviewed the scientific evidence on several possible options for CO_2 removal," it writes, "we conclude that these technologies offer only limited realistic potential to remove carbon from the atmosphere."[27] The Intergovernmental Panel on Climate Change (IPCC) isn't as blunt, but it too urges policymakers and the public not to rely on carbon dioxide removal. In its latest report on the physical science of climate change, the IPCC notes that "CO_2 removal technologies are unable, or not yet ready, to achieve the scale of removal that would be required to compensate for current levels of emissions, and most have undesired side effects."[28] And in its Special Report on 1.5°C (SR 1.5), it warns that "CDR deployed at scale is unproven, and reliance on such technology is a major risk in the ability to limit warming to 1.5°C."[29]

"Innovation" in Fossil-Fuel Propaganda

To be sure, this is not how fossil-fuel interests talk about CDR. Rather the opposite. With remarkable message discipline, as though repeating

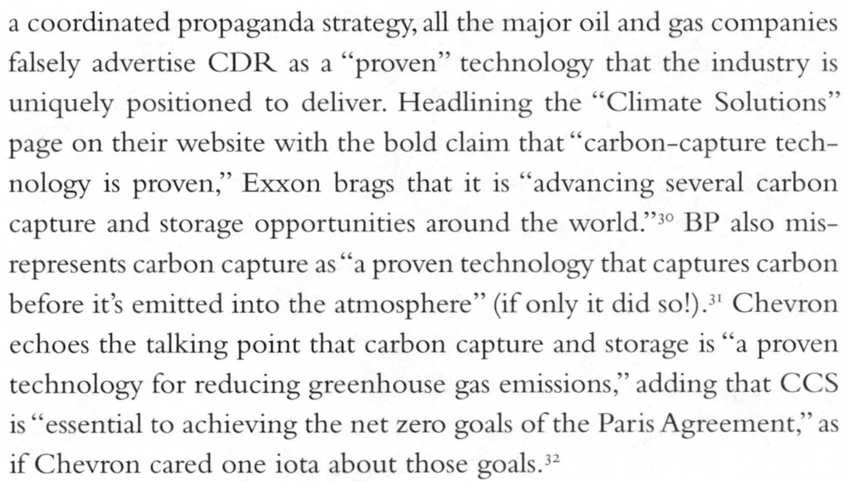

a coordinated propaganda strategy, all the major oil and gas companies falsely advertise CDR as a "proven" technology that the industry is uniquely positioned to deliver. Headlining the "Climate Solutions" page on their website with the bold claim that "carbon-capture technology is proven," Exxon brags that it is "advancing several carbon capture and storage opportunities around the world."[30] BP also misrepresents carbon capture as "a proven technology that captures carbon before it's emitted into the atmosphere" (if only it did so!).[31] Chevron echoes the talking point that carbon capture and storage is "a proven technology for reducing greenhouse gas emissions," adding that CCS is "essential to achieving the net zero goals of the Paris Agreement," as if Chevron cared one iota about those goals.[32]

To extend their license to operate, fossil-energy companies do certainly need to pretend that they care about the Paris Agreement, maintaining the fiction that they are not perfectly willing to roast the planet to a crisp so long as they can continue to profit from selling their products. That is why they all have announced their ambitions to achieve "net zero." But most of the plans behind these ambitions remain by and large limited to decarbonizing "operations," which is to say the extraction of oil and methane gas, rather than eliminating the pollution generated by burning that oil and gas itself.[33] Presumably in crafting these largely nominal net-zero ambitions, the industry's public-relations consultants felt confident that the public would skim over the details of the plans themselves.

When fossil-energy companies do promise to eliminate the emissions from the combustion of the oil and gas they sell, they also bury important details of their plans. Only in the fine print does it become clear that their promises rely, almost entirely, on carbon dioxide removal. BP's aim to become a net-zero fossil-energy company depends partly on emissions "reductions" and partly on emissions "removals" whose final amounts will be "determined in accordance with BP's methodologies."[34] Shell uses the same strategy of counting on self-counted "removals" to clean up the emissions of their products: when they announced their target to "become a net-zero emissions energy business by 2050," the company explained that "net-zero" means not just reducing emissions using point-source CCS, but also "capturing and storing any remaining emissions using technology," as if bio-energy with carbon capture and storage, or direct air capture, could indeed remove all the climate pollution generated by large-scale fossil-fuel use.[35]

Out of all the oil majors, Occidental Petroleum (or Oxy, as it's renamed itself) is really doubling down on the false promise of CDR at scale. In their disclosure materials, they come right out and say that their "pathway to achieve net zero" is efficient "oil, gas, and chemicals production coupled with industrial-scale carbon management solutions."[36] In 2023 Oxy even went so far as to acquire Carbon Engineering—a company that developed a liquid-solvent DAC process in the early 2000s—in order to brand itself as a carbon-management company, while also continuing on as (according to the acquisitions disclosure) "one of the largest oil producers in the US."[37] Oxy's new carbon-management division will fit in nicely with its core business. If it succeeds in developing new DAC projects, it will not even directly sequester the carbon dioxide it captures; rather Oxy plans to use that CO_2 for "enhanced oil recovery," a process that injects pressurized CO_2 into depleted wells in order to force out more oil. If, as they promise, direct air capture will "deliver [their] shareholders value through an improved drive for technology innovation," that innovation will only enhance the efficiency of the existing fossil-fuel system, not disrupt it—while at the same time gilding Oxy as a kind of climate leader. That is surely a shareholder/public relations win-win.

Occidental CEO Vicki Hollub appears regularly in the press to promote the story that her company is "leveraging [their] expertise in carbon management and storage to achieve net zero."[38] ExxonMobil echoes her messaging, announcing plans "to use its technology expertise, particularly in carbon capture" to "support net-zero ambitions."[39] This is a core messaging strategy of oil and gas companies, to paint themselves as the "experts" on carbon capture and carbon dioxide removal in order to seem not only benevolent, but also indispensable. And it's a good strategy. Oil and gas companies do have expertise in geological mapping, and oil and gas workers do know how to take fossil fuels out of the ground. Why wouldn't they know how to put the carbon back where it came from? Well, one reason would be that the operation of CDR technologies does not rely on expertise in fossil-fuel extraction or refining. Bioenergy with carbon capture and storage is largely an agricultural process, and direct air capture draws on techniques developed not in energy production but in the pulp and paper industries. And the sorbents and solvents in DAC would be produced by chemicals manufacturers, not fossil-fuel companies.[40] And the sites identified by geologists as most promising for carbon dioxide storage are not empty oil and gas wells, but saline aquifers and basalt rock.[41]

Oil and gas producers know this, of course. But their claim to expertise in CDR is merely a pretense anyway. They don't really expect to turn themselves into carbon management companies. In the 2023 World Energy Investment analysis, the International Energy Agency reported that a mere 4 percent of oil and gas companies' upstream investment went into lower-emission sources of energy or decarbonization technologies. Meanwhile their investment into new fossil-fuel extraction, already the majority of their expenditure, rose by 11 percent in 2022 and an estimated 7 percent in 2023.[42]

Oil and gas companies are also lobbying fiercely to prevent the United States from passing any climate policy that targets fossil energy. According to one non-governmental organization report, since the Paris Agreement was signed (the Paris Agreement whose goals they all claim to support), the five largest oil majors have spent $200 million per year lobbying against climate policy through their national trade association, the American Petroleum Institute (API).[43] In the summer of 2021 alone, the API spent almost half a million dollars to run hundreds of social-media ads attacking the budget reconciliation package in which Biden hoped to pass his climate policies. These ads, some of which directly attacked the bill's supporters in Congress, were viewed at least twenty-one million times.[44] In October 2021, the House Oversight Committee held a hearing on "Big Oil's Disinformation Campaign to Prevent Climate Action," in which the top executives of ExxonMobil, Chevron, BP America, and Shell were called to testify. The Committee asked these men and women whether, given their stated commitment to the Paris Agreement, and given the API's opposition to the goals of that Agreement, they would renounce their membership in the trade group. They all refused. Their current disinformation strategy is to spread the propaganda that they're the experts in carbon capture and carbon removal—carbon management innovators who can help the world meet the Paris climate targets—while, at the same time, their trade group works behind the scenes to prevent or repeal legislation that might in any way threaten their ongoing profits.

The Dangers of CDR Advocacy

Instead of countering this fossil-fuel propaganda, academic and corporate advocates for carbon removal technologies all too often echo it, creating an apparent consensus that, using the magic of innovation, it is possible to continue to burn lots of fossil fuels but still achieve net-zero emissions. It's hard to understand the motivations of these advocates, although perhaps it's significant that many of them have ties to the fossil-fuel industry. S. Julio Friedmann, one of America's leading experts on CDR, is a geologist and fellow at Columbia University's Center on Global Energy Policy who worked for ExxonMobil before becoming the Principal Deputy Assistant Secretary for the Office of Fossil Energy under the Obama administration. Holly Jean Buck, a highly prolific and influential sociologist who advocates using both carbon dioxide removal and solar geoengineering, began her career as a "geospatial technician for a remote sensing company." Armond Cohen—the founder of the Clean Air Task Force, a self-described "environmental group" which lobbies for carbon capture on Capitol

Hill—is a Brown- and Harvard-educated lawyer who formerly worked with the utilities industry.

The Clean Air Task Force echoes oil and gas companies outright, calling CCS a "proven" technology that "has been working safely and effectively for almost 50 years."[45] On her part Buck uses clever implicatures to create the false belief that carbon dioxide removal technologies are sure to work, headlining her website, for example, with this call and response: "The tech to draw carbon out of the atmosphere and restore ecosystems exists. What social organization do we need to get it done?" This question implies that the only things impeding CDR are contingent social relations, not hard planetary limits. But of course this is not true. To say "the tech exists" is to say nothing about its capacity for planetary-scale deployment. Human life in space also "exists," in that astronauts live on the International Space Station for months at a time. No-one believes this means that everyone on earth could move to space permanently. And rather than restoring ecosystems, CDR at planetary scale, as we have seen, looks likely to further degrade them. On his part, Friedmann repeats the disinformation that carbon dioxide removal can restore ecosystems. For example, citing the "unplanned geophysical experiment" of global heating in a journal article, Friedmann proclaims that "we've developed the technology to both stop the experiment and ultimately to undo it." "Cleaning our collective room may be unpleasant," Friedmann goes on, sympathetically, "but is ultimately necessary and is the work of climate restoration."[46]

This misleading word "restoration" resonates across CDR discourse. An advertorial placed in the environmental publication *Grist* by the carbon-removal start-up Climeworks, for instance, is duly entitled "Carbon capture: a critical tool in the climate restoration toolbox."[47] Yet the scientific evidence shows that, even if we could deploy enough carbon dioxide technology to cool the planet, cooling the planet with CDR would not "restore" the climate. The latest IPCC report explains that "if global net negative CO_2 emissions were to be achieved and be sustained, the global CO_2-induced surface temperature increase would be gradually reversed but other climate changes would continue in their current direction for decades to millennia."[48] "Some impacts," the report continues, "will cause release of additional greenhouse gases" and "some will be irreversible, even if global warming is reduced."[49] Sea level rise, in particular, would continue for at least several centuries, drowning coastal cities.[50] Nor is there a one-to-one relationship between emissions and removals: if CDR managed to achieve net negative emissions, the ocean would start off-gassing the extra carbon dioxide that it has absorbed from human pollution, requiring the world to remove two billion tons to achieve a reduction of one billion tons of atmospheric CO_2.[51]

These CDR advocates ignore or suppress this science, promising that climate restoration and decarbonized fossil fuels are largely doable, if only we would get it done. Buck, in particular, disseminates the view that we can run the fossil-fuel system cleanly, asserting in one article, for instance, that "with carbon capture and storage, zero-emission fuels and zero-emissions fossil-fueled electricity also becomes possible."[52] (Friedmann seems to believe this too, once quipping to a reporter: "It's convenient to say 'Keep it in the ground' . . . What I prefer to say is 'Keep it from the air.'")[53] Not content to promise that fossil fuels could be merely carbon neutral, Buck sometimes even goes further, claiming that "combining direct air capture with enhanced oil recovery could produce carbon-neutral or even carbon-negative fuel."[54] This is, of course, precisely the propaganda advanced by Hollub, the Occidental CEO, on her news-media tour to rebrand Oxy as a "carbon management company." Hollub once told ABC News that "We believe direct air capture is another way to also help the world because . . . you can put it anywhere you want to put it. So by combining it with oil reservoirs, that's the best of all worlds. To be able to provide the world with negative carbon oil, while also being able to meet the needs of the world to use oil."[55] Hollub, like Buck, uses the idea of direct air technology to say "we shouldn't kill fossil fuels"—or keep them in the ground—because "we've got to just figure out how to make fossil fuels emission-less"—or keep the carbon from the air. Astonishingly, Buck praises Oxy for their rebranding as a "carbon management company"—calling them one of the "early movers . . . anticipating markets for lower-carbon fossil fuels." She even argues that Oxy's PR campaign is "a step towards getting people ready to buy oil that will be carbon neutral by way of direct air capture."[56]

Buck epitomizes the way that CDR advocates often argue for the continuance of the fossil-fuel system even as they say that we need to end the general use of fossil fuels. Doublespeaking across and within her publications, in one article Buck claims that "the availability of decarbonized fossil fuels changes the politics around the idea of 'ending fossil fuels,' "because such decarbonized fossil fuels (whose existence she does not doubt) enable a "'reasonable' consensus . . . that fossil fuels will be part of the energy mix for the foreseeable future."[57] At the same time, Buck has written an entire book entitled *Ending Fossil Fuels*, where she argues that although a "cleaner fossil world is possible," fossil-fuel production should be phased out and the oil and gas companies nationalized and turned into "carbon removal companies."[58] Yet here too fossil fuels stay in the mix. For it turns out that being a "carbon removal company" doesn't mean (just) delivering negative emissions; it means continuing to supply fossil fuels *while* delivering negative emissions—producing some quantity of fossil fuels but taking that production back, as it were, by removing the carbon of these fuels from the atmosphere. In Buck's words: "Instead of asking 'How do we end fossil fuels?' we can ask 'What do we do with the fossil fuel industry?' If the answer is just 'put it out of business' that's a missed opportunity." A better plan, according to Buck, is for the government, having nationalized the industry, to legislate "a mandatory link between production and sequestration" requiring "extractors and importers of fossil fuels to permanently store a percentage of the CO2 generated by their products" which would "increase over time to 100 percent."[59] This plan will enable consumers to continue to choose fossil fuels even as they increasingly embrace zero-carbon technologies like heat pumps or bicycles. "The only thing that can provide those choices," she insists, "is public investment now in innovation."[60]

It matters that these CDR advocates are echoing the propaganda of the fossil-fuel industry. For one thing, journalists often repeat their talking points, or even quote them directly, and thereby end up inadvertently legitimizing climate disinformation. This happened especially after the 2020 election, when journalists on the climate and energy beats began to write a lot about CDR. It's not clear what instigated this surge of interest: perhaps their contacts at the Department of Energy started talking up the technology, or maybe the press agents of oil and gas companies started to pitch stories about those companies' plans to "advance innovative solutions for a lower-emission energy future," as ExxonMobil likes to say. Or perhaps journalists' interest in carbon removal had been piqued by Bill Gates and Elon Musk's declaring that they were going to fund research into direct air capture.[61] (Gates announced January 2021 that he would invest in DAC research through his Climate Innovation Fund. Journalists responded to this announcement with a spate of articles praising his foresight and talking up the potential of the technology. Not to be outdone, a month later Elon Musk announced he was offering a $100 million prize, which he called an "XPRIZE," to the best CDR proposal, and once again the news media went bananas.) Now, when journalists recognize that a topic has become "newsy," they often turn to experts to get up to speed—which they usually need to do very quickly, sometimes in a matter of days. But the problem with this method of doing research is that experts in technologies are often advocates for those technologies. As the environmental philosopher Dale Jamieson points out, writing about solar geoengineering: "A research program often creates a community of researchers that functions as an interest group promoting the development of the technology that they are investigating . . . The researchers are the experts and frequently hold out high hopes for a rosy future if their technology is developed."[62] Of course, journalists are usually well aware of the bias of their sources, as well as their own biases, but even so they will often try to manage these conflicts of interest by acknowledging a "debate" or "controversy" about the technologies they are actually describing from an advocate's point of view.

In the case of carbon dioxide removal, many climate and energy journalists turned to Friedmann, Buck, and the Clean Air Task Force for guidance and pull-quotes when they wrote on CCS and CDR. And that is unfortunate, because there are other CDR researchers who talk about these technologies in ways that are not misleading. One of the most prominent of these researchers, Jennifer Wilcox of the University of Pennsylvania, fully acknowledges that "there is an unsettling gap between the assumptions made about the potential of NETs . . . and what is currently known about their feasibility of operating at a global scale."[63] Wilcox is sometimes quoted as an expert by journalists—and she should be more often. For the picture of CDR that emerged in mainstream and even left-wing news media in the early 2020s was the one advanced by the advocates who expressed the fewest doubts about CDR's feasibility.

When the overconfidence of CDR advocates gets spread by the news media, it influences voters' feelings about climate change, and possibly dampens their willingness to phase out fossil fuels. At least one study in climate psychology has found that learning about carbon removal erodes support for climate policy by lessening the perceived threat of global heating.[64] And this psychological effect makes sense. As the novelist George Eliot observes in her masterpiece *Middlemarch*: "We all know the difficulty of carrying out a resolve when we secretly long that it may turn out to be unnecessary. In such states of mind the most incredulous person has a private leaning towards miracle."

If there's any constituency who seems to be leaning toward miracle, and quite publicly at that, it would be the policymakers in the US federal government, Republican and Democratic alike, who are hoping that by seeding the development of carbon capture and carbon dioxide removal, they can ensure that domestic fossil-energy production will continue to rise while American emissions fall to net zero.

Their unrealistic or perhaps uninformed hope makes them vulnerable to influence by fossil-fuel lobbyists—of which there are many. According to a recent NGO analysis, more than 240 groups, including Chevron, the American Petroleum Association, and the American Gas Association, disclosed lobbying on carbon capture in 2020 alone.[65] Accordingly, over four administrations Congress has given oil and gas companies generous incentives to scale up carbon capture and storage—and now they're doing the same for carbon dioxide removal as well.

Under the second Bush administration, Congress introduced a tax credit, called "45Q," which subsidizes coal, oil, and gas production or refining that uses carbon capture. Although Bush was a fossil-fuel supporter, President Obama made no move to repeal this tax credit when he took office, and his Interagency Task Force on Carbon Capture and Storage released a report in which it concluded that "CCS can play an important role in domestic greenhouse gas emissions reductions while preserving the option of using abundant domestic fossil energy resources."[66] Trump shared Obama's desire to nurture the domestic fossil-energy sector, and he also looked to CCS to clean up its emissions—often saying in public that he wanted "clean coal, really clean coal." Under the Trump administration, Congress expanded the 45Q tax credit to $50 per ton of carbon dioxide captured and stored. When it passed the Inflation Reduction Act (IRA) in the second year of Biden's presidency, a Democratic Congress increased this credit yet again to $85 per ton. It also increased the credit that 45Q afforded to enhanced oil recovery. It is not clear why enhanced oil recovery needed to be subsidized, of course, since the practice was already well established. (In 2022, for example, 90 percent of the CO_2 captured by CCS was used to produce more oil.)[67] But even after decades of the fossil-energy industry being subsidized to develop the carbon-capture technology they claim is already "proven," this Congress apparently believed the increased tax credit was worth enacting in the hope that, this time, CCS might actually work.

The 45Q tax credit was not the only subsidy for carbon capture passed by Biden's Democratic Congress. As part of its 2020 Covid-19 stimulus, it appropriated more than $6 billion for CCS research, development, or demonstration.[68] And it earmarked more than $10 billion for CCS and direct air capture in the 2021 Infrastructure Investment and Jobs Act—a law which Manchin later praised, incidentally, for investing "billions of dollars into clean energy technologies so we can continue to lead the world in reducing emissions through innovation."[69] Many of these new subsidies are being routed through the Department of Energy's Office of Fossil Energy, the office that, among other duties, approves new US fossil-fuel extraction. In 2021 this office was rebranded as the Office of Fossil Energy and Carbon Management. The erstwhile director of Carbon Capture Coalition, an international organization whose partners include Shell, Occidental, and Peabody Energy, among others, was nominated and confirmed as the head of that office. (The Carbon Capture Coalition had long lobbied Congress not just to expand the 45Q tax credit, but also to drop the capture standard for the credit entirely, so that fossil-energy producers and power plants could receive the subsidy even if they never captured a single ton of carbon dioxide.[70]) When the Coalition's director, Brad Crabtree, testified at his nomination hearing, he used the fossil-fuel communications script—calling CCS a "long proven" technology—and made the dubious claim under oath that its record overall has been "quite positive."[71] In 2022, another CDR advocate who uses fossil-fuel talking points was installed at the Office: Holly Jean Buck became a Management and Program Analyst at the FECM Division of Strategic Engagement, the networking arm of the Office of Fossil Energy and Carbon Management.

Fossil-energy interests are able to use carbon-capture propaganda to stay at the center of federal innovation policy because the Biden administration, like the Trump, Obama, and Bush administrations before it, is credulously hoping, it seems, that the industry will deploy CCS and DAC as climate solutions that sustain the fossil-fuel system. Biden's Secretary of Energy Jennifer Granholm told Bloomberg

TV flat out that "these kinds of technologies will help for the oil and gas sector to be able to ramp up production, but in a way that's clean."[72] In this statement, which reiterates that the Biden administration wants oil and gas production to rise, Granholm sounded a bit like Occidental's Hollub, who has said that her company's "pathway to achieve net zero" is "oil, gas, and chemicals production coupled with industrial-scale carbon management solutions." Perhaps, then, it might not be too surprising that in 2023, less than a week before they bought Carbon Engineering, Oxy was awarded an estimated $600 million grant by the Department of Energy's Office of Clean Energy Demonstrations for the construction of a direct air capture "hub" in South Texas. In the announcement of this award, Hollub thanked the Department of Energy's leadership and said that she would "look forward to [their] partnership to deploy this vital carbon removal technology at climate-relevant scale."[73] The announcement also noted that Oxy's hub would be designed to remove up to one million tons of CO_2 per year. Far from being "climate-relevant," however, this amount of removal would represent just fifteen minutes of annual global emissions. It's hard not to feel that American taxpayers should have gotten more for their $600 million.

How to Talk about CCS and CDR

When you talk about carbon removal, the most important thing is to guard against the false narrative that CDR is a miracle that can decarbonize fossil fuels and restore the climate, if only the world would deploy it wisely. False promises about CDR are propagandistic: they misrepresent reality to sustain an oppressive power—here the power of oil and gas companies—that is harming the world. Rather than rely on oil and gas companies to resolve the existential crisis they have done so much, and are still doing so much, to cause, the world must phase out fossil fuels. That is the core message.

However, when circulating this message, you will need to be prepared to address one particularly sticky objection: that the science says we need large amounts of CDR if we are to meet the goals of the Paris Agreement and halt global heating at 1.5° or well below 2°C. This objection is very challenging because it is a claim underwritten by the IPCC itself.

How can this be? Well, in addition to reporting on the physical science of climate change, and on the ways that global heating will damage the planet, the IPCC also reports on the economics of halting warming. Working Group III, the branch of the IPCC that writes these reports, is staffed not by earth-system scientists, but by engineers, political and social scientists, and economists. As part of their work for the IPCC, these researchers aggregate and summarize mitigation pathways—which are, as we saw in "Alarmist," models that game out how to achieve net-zero emissions in time to halt global warming at particular temperature targets. The vast majority of mitigation pathways in the literature model massive amounts of CDR in order to halt warming at 1.5° or even 2°C in 2100. Indeed, the pathways discussed in recent IPCC reports model up to one trillion tons of carbon dioxide removed from the atmosphere by 2100.[74]

It may be worth pausing here for a moment. You may feel incredulous, shocked, or dismayed by this news. Perhaps I should confess that I felt devastated when I learned that the models centered by the IPCC place such a heavy bet on technologies that the science itself warns are infeasible at planetary scales. Is there no institution that is not complicit?

Here are three things to keep in mind. First: the IPCC scientific reports are very clear that CDR cannot be relied on at massive scales. Remember that in SR 1.5 the IPCC warned that "CDR deployed at scale is unproven and reliance on such technology is a major risk in the ability to limit warming to 1.5°C." And in its 2021 report the scientific working group warned that "CO_2 removal technologies are unable, or not yet ready, to achieve the scale of removal that would be required" and "most have undesired side effects." But of course this begs the question of how Working Group III can recommend removing a trillion tons of carbon dioxide with technologies unable, or not yet ready, to achieve the scale of removal that would be required.

Well, the second thing to keep in mind is these models are not recommendations. They are, as the IPCC itself affirms in its charter, meant to be "policy relevant, not policy prescriptive."[75] Indeed, many scientists and economists who responded to a recent survey of scenario modelers were eager to clarify that their work is "explorative" rather than "predictive" and should not be taken to prescribe "the eventual deployment of BECCS or negative emissions."[76] Detlef van Vuuren, a scientist and modeler at the Netherlands Environmental Assessment Agency and one of the most influential members of the Integrated Assessment Modeling Consortium, once told a reporter, with some impatience, that modeling is "just an element, a tool to explore different trajectories on the basis of the knowledge we have today and to see what kind of things we might encounter."[77]

And this tool is designed to unlock a very limited box. For the models at issue (and this is the third thing to know) are "targeted optimization models," specifically designed to find technology deployments that meet temperature targets by 2100 at the lowest possible cost. So of course they come up with strange results. Their assumptions practically guarantee it.

If the phrase "optimization models" reminds you of William Nordhaus, whom we met in "Cost," that's because Nordhaus' cost-benefit model is an analogue to these mitigation pathways. As Nordhaus' model finds that continuing to use fossil fuels, and allowing 3°C of heating, creates the "optimal" balance between the costs of climate policy and the costs of climate damages, so do these "targeted optimization models" find that continuing to use fossil fuels, and removing massive amounts of their emissions later in the century, is the optimal way to reach net zero emissions by 2100. But this finding relies entirely on how the models are designed. First of all, these mitigation pathways model neither climate damages nor loss of economic productivity from climate impacts; rather, they assume that the economy will always run at full capacity—perfectly efficient and frictionless—and that global GDP will grow at a constant rate over the whole century.[78] This assumption allows them to assume, in turn, that the world will be much richer after 2050, and thus that it will be relatively cheaper to deploy CDR later than to prevent more global heating now. They disregard any uncertainty about this projection—or about CDR itself. Instead, assuming "perfect foresight," they stipulate that CDR is definitely going to work at scale and will never be subject to climate damages (such as fire or flood destroying bioenergy crops, for example). That means the models never price in the risk that CDR might fail.[79] Nor do they price the risks that planetary-scale CDR might pose to agriculture, energy, materials demand, or any other component of the economy.[80] Finally, these models hugely overestimate the price of renewable energy, which, as we have seen, has dropped so precipitously that solar is now the cheapest electricity in history. When renewables are priced accurately in mitigation pathways, the value of CDR drops up to 70 percent, depending on the energy-system sector under consideration.[81] Joeri Rogelj, the coordinating lead author of the chapter on mitigation pathways in SR 1.5, explains: "the perceived linkage between end-of-century outcomes and the amount of late-century net negative emissions is not robust; instead it is to a large degree driven by the design characteristics that underlie the cohort of scenarios that is currently available in the literature."[82] To sum it all up: yes, the world will need a limited amount of CDR to remove truly essential emissions, but the models that say the world may need up to a trillion tons of removal by 2100 are coming to that conclusion because their assumptions are faulty.

When you talk about CDR, it will be useful to lift up the voices of scientists who contest the dominant models' economic assumptions. Already in 2016, the climate scientists Glen Peters, whom we met in "Alarmist," and Kevin Anderson condemned the "unjust and high-stakes gamble" of relying on widespread carbon removal in modeled mitigation pathways, and growing numbers of researchers are amplifying their call to end the modeling of continued fossil-fuel production and consumption.[83] In 2023, David Ho, an oceanographer who served as an academic reviewer for Musk's CDR "XPRIZE" competition, wrote in the journal *Nature*: "We have to shift the narrative as a matter of urgency . . . We must stop talking about deploying CDR as a solution today, when emissions remain high."[84] Accordingly, researchers are designing mitigation scenarios to explore how to decarbonize the global economy without overreliance on carbon dioxide removal. In one recent study, Rogelj finds that simply constraining warming, rather than trying to find the most "cost-effective" pathway to a temperature target in 2100, enables the model to achieve net zero in 2050 and halt warming at 1.6°C without using unrealistic amounts of CDR to cool down the planet.[85]

Every tenth of a degree matters. The world is likely to breach a sustained global average of 1.5°C warming sometime between 2026 and 2042 if emissions are not rapidly slashed.[86] If that happens, though, the next target will be 1.6° of warming, and then 1.7° if need be. But the world can still halt warming below 2°C without having to remove a trillion tons of carbon dioxide from the atmosphere. To do that, however, fossil fuels will need to be phased out—and as quickly as possible. (Governments must not try to gloss over this reality using oily words like "abatement.") Instead of looking to the fossil-energy companies to deliver CDR, imagining absurdly that they're going to become the garbage collectors of their own pollution, policymakers should be figuring out how to wind down coal, oil, and gas as they scale up renewable energy. Further, they should be developing robust innovation policies that seed and nurture the zero-carbon technologies and processes for concrete, steel, transport, and agriculture that the world will need if it is to transform industrial production and halt global heating. It is time for everybody to throw off the false hope that the fossil-energy economy can stay largely the same but be somehow retrofitted to become safe for living beings. As we shall see in the next and final chapter, "Resilience," ending the age of fossil fuels requires not merely retrofitting, tweaking, or shoring up what we have now, but transforming the world and—with creative imagination and a commitment to justice—inventing something new.

6

Resilience

The last refuge of the false solutionists is the language of "adaptation" and "resilience."

—Michael Mann, University of Pennsylvania (*The New Climate War*)

I advocate for anti-resilience.

—Shalanda Baker, Northeastern University ("Anti-Resilience: A Roadmap for Transformational Justice within the Energy System")

Only an urgent system-wide transformation can avoid climate disaster.

—The United Nations Environment Programme (*Emissions Gap Report 2022*)

In the language of climate politics, the term "resilience" describes the qualities people, communities, and nations should try to cultivate in order to survive climate change. These qualities enable the capacity of "bouncing back and returning to a previous state after a disturbance," as the Intergovernmental Panel on Climate Change (IPCC) put it in its 2022 report "Climate Change: Impacts, Adaptation, and Vulnerability."[1] The cultivation of resilience is so closely tied to the process of adaptation in climate discourse that most researchers and institutions use the words "resilience" and "adaptation" interchangeably. So, when in 2015 the Paris Agreement established a "global goal on adaptation," it portrayed this goal as "strengthening resilience and reducing vulnerability to climate change."[2] In its 2019 "Action Plan on Climate Change Adaptation and Resilience," the World Bank laid out a program for helping poorer nations not only "manage climate risk," but also "prepare for disruptions, recover from shocks, and grow from . . . disruptive experience."[3] And President Biden's National Climate Task Force included among its "bold steps to strengthen the nation's resilience" various funding streams to harden and protect the nation's built environment, such as re-establishing federal standards that will reduce flood risk, strengthening response capacity with more federal firefighters, and providing financial and technical assistance to drought-stricken communities. The idea of "resilience" encompasses all these measures. Indeed, in its public-facing materials the Task Force doesn't mention "adaptation" at all.[4]

But it is misleading to conflate adaptation and resilience. Conflating adaptation and resilience misrepresents the physical reality of the climate crisis. As we have seen throughout this book, if we do not force decision-makers to phase out fossil fuels and eliminate emissions, climate change will cease to be a series of discrete disasters from which communities can just "bounce back." Climate change will remake the face of the world, destroying the functional connections between human and ecological systems on which civilization depends. And figuring adaptation as resilience is misleading in a second way: it suggests that the goal of adaptation policy should be to shore up the world's dominant systems—the very systems causing the climate crisis in the first place—while extending their reach into the Global South under the guise of sustainable development. In this respect, not only does the goal of resilience perpetuate systems of racial and economic injustice, it also works against phasing out fossil fuels, eliminating emissions, and actually halting global heating—ultimately harming everyone. Indeed, resilience discourse functions as the sort of propaganda that "irrationally closes off certain options that should be considered," as the Yale University philosopher Jason Stanley puts it in his book *How Propaganda Works*.[5]

People and institutions at the center of global power vaunt the idea of resilience in part because it elides the urgent need for systemic change. This is why the term "resilience" should be replaced by the word "transformation." For it is the transformation of our systems, and ourselves, that will in the end preserve a livable future.

Resilience and the Persistence of the Fossil-Energy System

The term "resilience" entered climate discourse from the 1970s' work of the ecological economist C. S. Holling. Holling defined resilience as measuring "the persistence of systems and their ability to absorb change and disturbance and still maintain the same relationships between populations or state variables."[6] Resilience is a very important quality for ecosystems to possess. Life would be endangered if forests, marine food chains, or pollinator populations, for example, could be pushed past their tipping points with just a little nudge. Human communities also need resilience to survive extreme weather, so the word has value insofar as it captures "the ability to prepare and plan for, absorb, recover from, and more successfully adapt to adverse events," as the National Academies of Sciences, Engineering, and Medicine put it in a 2012 report on disaster preparedness.[7] Measuring community resilience can help policymakers prioritize investments, allocate limited resources, and target the most effective programs to mitigate the effects of both discrete adverse events like hurricanes and ongoing stressors like droughts. But problems arise when the concept of resilience migrates away from the spheres of ecology or disaster preparedness and becomes the goal of climate policy meant to shape the holistic relationship between human and ecological systems under threat by global heating.

The first problem is that the term "resilience" misprizes the climate crisis. It implies that the harms of global heating will not include overall climatic breakdown, but only discrete extreme weather events, occasional and temporary, that will disrupt a world that is otherwise normal. But as we have seen throughout this book, if we, alarmed people everywhere, do not force decision-makers to phase out fossil energy—and to establish economies well-integrated with ecological systems—climate change will not remain a crisis of more frequent and severe extreme weather alone. It will mean ongoing heat that will make parts of our planet uninhabitable, the drying up of water sources that once supplied drinking water for billions of people, the inundations of coastal regions where billions of other people live, the decimation of animal and plant agriculture in breadbaskets across the world, and the resultant unraveling of anything resembling a normal economy or a normal life. Nor will extreme weather remain as Americans have known it so far—a California wildfire in July, say, followed by a midwestern flood in August, capped by a Florida hurricane sometime in September or October. Extreme weather disasters will happen simultaneously, in multiple regions of America at once, repeatedly—and potentially before communities have recovered from the last disaster.[8]

Even assessing future extreme weather events in themselves, it is naive to imagine that they would recede only to leave the ecological conditions for a full recovery in place. The 2022 floods that devastated Pakistan, killing 1,700 people and throwing millions more into poverty and homelessness, exerted so much force on the land that they caused some rivers to change their courses.[9] The lake that has given Salt Lake City its name is drying up, due to climate-change intensified drought. If it evaporates completely, the air around it will occasionally turn poisonous: the lake bed contains high levels of arsenic and as more of its surface becomes exposed, wind storms will carry that arsenic into the lungs of nearby residents.[10] The metro area containing nearly 40 percent of Utah's population could become a ghost town. Sea level rise also destroys cities permanently—once coastlines are swallowed by the oceans, or rendered uninhabitable by encroaching salt water, they will be gone effectively forever.[11] The coral reefs that support much of the marine food chain cannot be resurrected once they've fallen extinct. As we have seen, permanent changes like these mean that people will need to move away from where they were living in order to find safe homes and sources of income and food. This migration is also a form of adaptation—a kind of adaptation not well captured by the term "resilience." And even this kind of adaptation will find its limits. If the tropics become uninhabitable, where will the billions of people who live in that region actually go? As the IPCC warned in its 2022 report: "with increasing global warming, losses and damages will increase and additional human and natural systems will reach adaptation limits."[12] Or as the climate scientist Michael Mann has written, more bluntly: "there is no amount of resilience or adaptation that will be adequate if we fail to get off fossil fuels."[13]

As it distorts our view of possible climate futures and the real character of adaptation and its limits, the term "resilience" also obscures the socioeconomic causes of the climate crisis. It offers what the scientists Danny MacKinnon and Kate Driscoll Derickson have called an "apolitical ecology" that "not only privileges established social structures, which are often shaped by unequal power relations and injustice," but also "closes off wider questions of progressive social change" which create the "transformation" of "established 'systems.'"[14] By framing the goal of adaptation policy as building the capacity of human systems to bounce back and return to their previous state after a disturbance, the word "resilience" implies that the previous state of those systems was desirable to begin with. This implication silently glosses over how the current distribution of power across people and institutions—what Holling might call "relationships between populations or state variables"—is sustaining the kind of economy that leads to global heating. Critiquing the use of scientific concepts to explain historical phenomena, environmental humanities scholar Ashley Dawson has argued that policy meant simply to produce resilience "fails to question the political conditions" that give rise to the harms of climate change, assessing them "as if they were natural," and thereby enabling policymakers to attempt to mitigate the threat of extreme weather "without fundamentally transforming the conditions that give rise to these crises."[15] This is why highlighting the term "resilience" can even enable right-wing policymakers to adopt the pretense of taking climate action while never linking weather threats to the fossil-fuel economy, which these policymakers, of course, continue to support.

Republicans in regions particularly vulnerable to climate damages often use the language of resilience in this propagandistic way. Florida Senator Marco Rubio and Governor Ron DeSantis especially favor this tactic. In one *Washington Times* commentary, Rubio acknowledges that Florida is "uniquely impacted by the climate," and he touts his own environmental record ("far from denying the value of climate-conscious efforts, I have led them for years"). But it turns out that his record consists not of supporting any policy that might lower greenhouse gas emissions, but only of making "improvements to Florida's coastal resilience." Indeed, Rubio even suggests working toward resilience *requires* the use of fossil fuels: "people who are serious about increasing America's climate resilience agree that we cannot operate on solar and wind alone," he claims, falsely, "because in their current state, those technologies cannot supply enough power to keep our nation running." This is a very canny claim. It activates the fear, stoked by a combination of ignorance and oil and gas company advertising, that because wind and solar are variable they are not reliable. But the truth is that solar power and storage actually maintains grid reliability in extreme weather. In both the 2022 and 2023 heat waves, Texan grid operators needed to pause fossil-energy generation so that coal and gas plants could cool down; what kept the lights on and life-saving air conditioning units humming was solar power both generated directly and stored in batteries.[16] Dozens of scholars at European universities and American institutions like Stanford and Princeton have modeled how to supply all of our nation's power with 100 percent renewable energy.[17] Yet Rubio's commentary ignores all these facts and, for good measure, spreads other denialist canards, such as the absurd claim that Democratic climate policies would cause worldwide famine (presumably because those policies would supposedly outlaw nitrogen-based fertilizer?)—which means, according to Rubio, that "'ending' fossil fuels would kill many more people than global warming." This claim rehearses the sort of upside-down propaganda that George Orwell examined in his novel *1984*, where the slogans of Big Brother's totalitarian Party are "war is peace; freedom is slavery; ignorance is strength." As we saw in the "Introduction," Vivek Ramaswamy also made use of this kind of propaganda in a Republican primary debate when he said that "the reality is that more people are dying of bad climate change policies than they are of actual climate change." The idea is to create the belief that the world must cling to fossil fuels for survival—never mind the dangers of global warming, which will be held off, in Rubio's telling, by "a comprehensive, realistic approach" to "climate resilience."[18]

More of a culture-warrior than Rubio, Ron DeSantis originally staked his failed 2024 presidential campaign on a crusade against what he called "woke ideology," including the climate crisis which he dismisses as nothing more than "left-wing stuff."[19] Yet DeSantis too has been bullish on "resilience." Two years after winning the Governorship, he used money from the American Rescue Plan, the $1.9 trillion Covid-19 relief act passed under the Biden administration, to establish what he called the "Resilient Florida Program," which distributes grants to help "prepare communities for the impacts of sea level rise, intensified storms and flooding," according to the Florida Department of Environmental Protection.[20] That funding has been inequitably distributed, however. The investigative journalist Craig Pittman found that most funds administered by the program have helped to protect the real estate in which some of DeSantis' staunchest campaign contributors have invested. Other than that, the Resilient Florida Program has done little to prepare Florida for extreme weather, leaving pre-climate-era building codes in place while the State government has continued to approve new development in areas where evacuation times already exceed state regulations.[21] Still, every six months or so, DeSantis announces the award or disbursement of more funds from the program, maintaining an image as a climate-responsive Republican—an image that is burnished when centrist conservation groups like The Nature Conservancy praise DeSantis as an environmental leader for his commitment to resilience.[22]

Of course, even as he distributed resiliency grants, Governor DeSantis continued to throw the full weight of his executive office behind fossil fuels. In 2021 he outlawed the passing of any ordinance that might help phase out those fuels, signing a bill that prohibits the enactment or enforcement of any "resolution, ordinance, rule, code, policy" or "action" that bans or even merely restricts any particular source of energy from being supplied to Florida ratepayers by a utility or transmission company.[23] This law targets, among other things, bans on methane gas in new construction, even though fully electrified options for heating, cooling, and cooking have already entered the market and are being robustly supported by the Inflation Reduction Act. From his platform as a member of Florida's State Board of Administration, DeSantis also announced that the managers of Florida public pension funds would no longer be allowed to consider environmental, social, or governance (ESG) principles in their investments. (ESG principles help pension managers to guide investments out of fossil energy and into climate-safe companies and funds.) DeSantis mandated that his state's public fund managers must now only assess "pecuniary considerations."[24]

Here DeSantis was participating in a larger Republican strategy. In August 2022, nineteen Republican Attorneys General wrote to Larry Fink, the CEO of BlackRock, the world's largest asset manager, to insinuate that they might pursue legal action against his company for breaking what they called "state laws requiring a sole focus on financial return" in service of a "climate agenda."[25] This was more upside-down nonsense. Considering BlackRock was holding $260 billion worth of fossil-fuel-company equity at the time, it was hardly the case that it was advancing a "climate agenda." What it was actually doing was diversifying its portfolio. Indeed, ESG investing is very much to the benefit of what DeSantis called "pecuniary considerations." A study out of Canada's University of Waterloo found that by 2022 the total cumulative value of six major US public pension funds in Alaska, California, New York State, Oregon, and Wisconsin would have been about 13 percent higher had these funds divested from fossil-fuel holdings a decade before.[26] But the goal of attacking asset managers for investing in ESG funds is not really to secure the retirement accounts held in public pensions. Its goal is to block any capital from flowing into climate investments. That DeSantis can participate in this kind of state intimidation in support of the oil and gas industry while he touts his Florida Resilience Program speaks directly to the way the rhetoric of "resilience" all too easily dovetails with the continuance of the fossil-fuel economy.

From Resilience to Transformation

If the term "resilience" can be readily co-opted by fossil-energy interests, the word "transformation" may put up more resistance to appropriation. It is hard to imagine fossil-fuel interests calling for the "transformation" of the world. That is one reason you should start talking about transformation instead of resilience. The other, and more important reason is that resolving the climate crisis will in fact require the transformation of the world. The United Nations Environment Programme has outlined a comprehensive plan to "accelerate transformations in electricity supply, industry, transportation, and buildings"—essentially the entire material foundation of the economy—by mobilizing international organizations, national and subnational governments, private and development banks, businesses, and individual citizens.[27] And the IPCC has signaled that these material changes will necessarily be driven by and produce further transformations in culture: a "sustainable world," it tells us, "involves fundamental changes to how society functions, including changes to underlying values, worldviews, [and] ideologies."[28]

It may seem like an overwhelming task to change everything, everywhere, all at once. So let's just start somewhere. Let's focus our minds on what Shalanda Baker, the Director of the Office of Economic Impact and Diversity at the Department of Energy, has called "anti-resilience." Baker defines anti-resilience as the action of resisting "the systemic violence enacted upon communities of color and the poor in the name of energy." Anti-resilience, Baker elaborates, motivates "a politics of anti-racism and anti-oppression that exposes the roots of structural inequality" and "illuminates the path for system transformation."[29] Why stand for anti-resilience by connecting climate change to anti-racism and anti-oppression? Well, anti-racism and anti-oppression are the foundation of climate justice—the moral imperative to accept responsibility for and repair the reality that the people who have been most harmed by the pollution of fossil fuels, and who are now suffering first and worst from the harms of global heating, are the people of color in America and across the world who have emitted the least CO_2 and done the least in all other ways to cause the climate to break down. Yet climate justice should be the first principle in undoing the climate crisis not only because it is morally right. It should guide climate action because the world will likely fail to halt global heating if decision- and policymakers continue to believe, falsely, that the climate can support sacrifice zones where the externalities of the fossil-fuel capital will remain confined. I hope this book has shown that the belief that some planetary zones and people will remain decoupled from environmental externalities, even if the world continues to use fossil fuels, is an ideological illusion.

It is also important to understand that subsidizing the build-out of renewable energy nationally but simultaneously ignoring Black communities' calls to close fossil-energy power plants and refineries in their communities will simply construct an all-of-the-above energy system in which fossil fuels persist, and will thereby fail to decarbonize the American economy. To wit: the Biden administration had promised that 40 percent of "the benefits" of its climate policies would flow to disadvantaged communities of color, but then it characterized funds given to coal and gas plant operators to retrofit their facilities with carbon capture as environmental-justice money, simply because those facilities would be sited where Black communities live.[30] The racism underlying this misrepresentation of a fossil-fuel subsidy as environmental-justice money becomes all the more obvious in light of the fact that Biden's own environmental-justice task force recommended against counting carbon capture as a climate solution.[31] As we have seen, carbon capture does not capture anything near the majority of combustion emissions, making it an entirely inappropriate solution for the power sector. Even though the Supreme Court has constrained the Environmental Protection Agency's capacity to regulate carbon pollution from power plants, the Biden administration could still mandate that polluters themselves should pay for carbon-capture retrofits while imposing price caps on electricity generation to ensure their costs would not be passed on to ratepayers. But such policies have never been floated for public debate. When policymakers ignore the voices of Black people and people of color in their attempts to "abate" fossil fuels, African Americans are the people most harmed. But everybody loses.

Further, insofar as the undoing of climate change will require keeping Republicans out of the majority, a fundamental element of the political work of halting global heating will be building and sustaining a multi-racial coalition of people who are alarmed about the climate crisis and/or who stand to benefit immediately from the new jobs and the rise in their real incomes that will be won by phasing out fossil energy. This coalition must be centered on Black people and people of color. When African Americans turn out at the polls, Democrats win. As Rhiana Gunn-Wright, an architect of the Green New Deal, has written:

> from 2018 to 2022, the climate movement became more powerful and unified than ever because we coalesced around a vision of green transition that included racial justice and broad social welfare policies. When I helped design the Green New Deal, I was trying to design policies that served Black people. And that proposal helped galvanize millions

> of people to care about climate change; they created local coalitions, passed municipal and state versions of Green New Deal policies, and eventually pushed climate to the top of the Democratic agenda. That, in turn, opened up the space to leverage the political opportunity created by the Green New Deal into meaningful federal action.[32]

Without the focus on improving Black lives, in other words, the climate movement will fragment. Just as crucially, without an anti-racist, anti-resilience framework, the immediate benefits of climate policies will be distributed to constituencies, like college-educated, homeowning white Democrats, who support those policies anyway, or they will flow toward working-class Republicans who are averse to supporting climate policy due to their partisan commitments, even if their employment prospects might improve from the creation of new, clean-energy industries in their states. But improving Black lives—cleaning up pollution in African-American communities, expanding economic opportunities for young people of color in new clean-energy manufacturing jobs—will fortify the base of the Democratic party and make it unshakable.

Of course, not everyone agrees with this view. Some voices in the climate movement argue that jobs created for white people by Inflation Reduction Act investments in red states will indeed create a new, bipartisan coalition that supports clean energy. Other voices say that worrying about coalition-building is unnecessary—governments should just put a tax on carbon and let the market allocate winners and losers once climate damages are fully priced in. Yet it seems clear that people of all political persuasions vote not just to maximize their economic interests, but to preserve their own sense of their identities and their social positions, which will make overcoming right-wing climate denial with jobs more challenging than it may seem. And even if carbon taxes were a viable solution, it would take many decades, possibly even a century or more—time we do not have—for the invisible hand to create historical change. The United States Army used horses to pull its artillery as late as the 1940s. It was only the federal government's unprecedented mobilization of the productive forces of the economy that transformed America into the pre-eminent industrial powerhouse that it became by the end of World War II.[33] That is the kind of mobilization that the world needs, in as many governments as possible, collaborating to force the transformation of global systems despite authoritarian fossil-fuel resistance. This is why the *sine qua non* of saving the world is building a mass movement to elect and support candidates who will commit their power and the power of their offices to this mobilization. As Gunn-Wright has said: "We need a sustained mass movement along with mass organizations for climate. And the only way there is with—not over—Black people."[34] Anti-resilience, anti-racism, anti-oppression—the unequivocal goal of transforming our current systems, in all their injustices—is not a distraction from the work of decarbonization. It is the core of that work.

This pronouncement may seem utopian. To sustain a coalition that encompasses coastal Democrats alarmed about climate change, Black people and people of color fighting toxic pollution in their communities, young people furious about their future being stolen by the rich, and, possibly, climate tech entrepreneurs—and then to have this coalition force the US government to treat climate like an emergency and mobilize on anti-racist principles? That seems hard. But it may be doable.

As we saw in "Alarmist," the political scientists Erica Chenoweth and Maria Stephan have studied 323 violent and non-violent campaigns that occurred between 1900 and 2006 worldwide; they found that every campaign that mobilized at least 3.5 percent of the population in sustained non-violent protest achieved its stated objectives within two years of the campaign's end.[35] Americans are not yet mobilized for sustained civil disobedience, by any means, but nearly 3.5 percent of them *already* believe climate change is the most important issue facing America today—a February 2023 Gallup poll reported that number as high as 3 percent.[36] Nor is this 3 percent merely a radical fringe. An August 2023 NPR/PBS NewsHour/Marist poll found that a majority of Americans—53 percent—felt that addressing climate change should be given priority *even* at the risk of slowing the economy. That majority included 80 percent of Democrats and 54 percent of independents.[37] Again, these people are not yet mobilized, but they are there.

This is why one of the most powerful climate actions you can take is to join or donate to organizations engaging in multi-racial, cross-class organizing around energy transformation and climate justice issues. And if organizing is not your cup of tea? That's totally fine. A movement meant to transform the world must embrace strategic and tactical diversity. So ask yourself: who am I? What are my talents and resources? What do I like doing? How can I leverage my position in the places I live and work, in whatever my social circle may be, to help phase out fossil fuels and resolve the climate crisis. The goal, here, is to transform *yourself* into someone who rises to this epochal moment, takes on the climate crisis intellectually and emotionally, and does what they can to help.

To do this you will need courage, of course. But you will also need faith. You will need to hold unseen possibilities in your mind. The Black feminist and activist Angela Davis once told a lecture hall: "you have to act as if it were possible to radically transform the world. And you have to do it all the time."[38] This willingness to act as if climate justice were possible, and to do so all the time, can be challenging. It can be socially alienating in a world where climate silence is the norm. But committing to ending the fossil-fuel system does something profound and personally valuable: it gives your life meaning. It turns you into living proof that the fossil-fuel economy is not an expression of human nature, but a contingent way to distribute power. For here you are, a human being, fighting to distribute power differently.

If this feels too woo-woo, then don't think about faith. Instead, try to cultivate humility and embrace intellectual uncertainty. Admit that you don't know, and can't know, what is going to happen—so you might as well proceed as if the world can be otherwise. The author and activist Rebecca Solnit has written that "hope is not about what we expect," but an "embrace of the essential unknowability of the world":

> I believe in hope as an act of defiance, or rather as the foundation for an ongoing series of acts of defiance, those acts necessary to bring about some of what we hope for while we live by principle in the meantime. There is no alternative, except surrender. And surrender not only abandons the future, it abandons the soul.[39]

In the end it is your soul to whom you are accountable. When you look back over the arc of your life, how you touched other people and the planet you're leaving behind, what will your soul say to you? What will be the meaning of your life on this planet? After your death, the carbon dioxide you put into the atmosphere will remain there for thousands of years, even as the particles that made your body will unwind and the waves of energy that animated you will go on to animate your descendants. How will your descendants live? How will they remember you? (If you're a parent: how will your own children remember you?) What you do to help resolve the climate crisis will shape your immortality in the memories and the struggles and the triumphs of the young people everywhere who are alive right now.

Climate change may well inspire a reckoning for you about who you are and what it means to be human. Let it do so. But always remember: this is a battle against the forces of destruction to save something of this achingly beautiful, utterly miraculous world for our children. The fossil-fuel industry and the governments that support it are literally colluding to stop you from transforming the world. They are trying to maintain the fossil-fuel economy. As for me—and as for you, here with me at the end of this journey, this book—I will say: *we* are against them, and we are going to fight for dear life.

After Words: Walking the Talk

If you're not used to doing it, talking about climate change can be very difficult. It can feel risky or socially awkward, as if by bringing up the climate crisis you're betraying social norms against disturbing other people. Well, you are. But that's ok. There is a time for speaking out, and that time is now. As I said in the first chapter of this book, one of the most powerful tools you possess is your voice—talking about the dangers of global heating and the people who are still committed to promoting fossil fuels is a very effective way to disrupt systems of denial and complacency, which depend on codes of silence to seem normal. According to researchers at Yale University, around 64 percent of Americans never discuss climate change in their social circles.[1] That statistic needs to change and fast.

At the same time, how you talk to people matters, and you cannot talk to every person in exactly the same way. When you are having a conversation or an interaction on social media with a person or a group of people, you will need to listen very actively and compassionately to your interlocutors. Repeat back to them what you hear, kindly, and make sure you understand what they're saying. Try to hear the feelings and the beliefs *behind* what they're saying. Are they disengaged from the climate crisis because they're overwhelmed with financial or personal worries? Do they feel conflicting loyalties because a family member watches a lot of Fox News? Do they understand that climate change is dangerous but know nothing about solutions? You will need to improvise and shape the information you share to try to address both the anxieties and the blind spots plaguing the people you're speaking to. And you must be honest with yourself about how much your conversational partner trusts and respects you. You will need to be trusted and respected for your message have power.

No one conversation—or op-ed, or letter to the editor, or social media post, or speech at a dinner party or community meeting—will transform someone into a climate activist. If your audience is one of the alarmed and you have an opportunity for deep engagement using the three-part messaging I discussed in Alarmist (inspiring fear, outrage, and desire for change) you might fire up their determination and motivate them to take new actions. But when you talk to people who aren't engaged with climate change yet your goal should be simply to plant a seed that might grow into something more substantial with time. If you are speaking to someone who dismisses climate change as a hoax, do not bother to debate them unless there are people observing your exchange who might learn something from your debunking false statements. If you encounter someone who's relatively uninformed about the climate crisis, or indifferent but open to discussing the problem, it will be enough if you connect the extreme weather you will both surely be experiencing to the global heating caused by fossil fuels. But don't belabor the point. Immediately pivot to touting clean energy, talking up how much you love your e-bike or electric vehicle, if you have one of those, or how much money you're saving having requested community solar through your utility or having put solar panels on your roof. If you've made it a biweekly habit to call your elected representatives and tell them they need to prioritize phasing down fossil energy in order to keep your vote, talk about that. But keep it light, keep it easy, and don't scare the horses. In these conversations you're just planting seeds.

On the other hand, if someone comes to you to talk about how frightened and sad they are about what's happening, or about how governments are still committed to supporting fossil energy, meet them where they are. They will benefit from your validation and compassion. Once they feel less alone in their experience, then you can start talking about the truths I discussed in Resilience: the need to embrace uncertainty, to cultivate the faith that good things might happen that they cannot anticipate yet, to imagine their legacy as people who fought for the world and the people they love. That might be a conversation you will need to have with them—and perhaps with yourself—more than once.

It is likely that if you are someone who has read all the way through this book (or even just picked it up and flipped to the end), the people you'll be talking to most often are Democrats who believe that climate change is real, but whose thinking about the crisis has been indelibly influenced by the fossil-energy propaganda that shapes the dominant consensus. Indeed, as we saw in the Introduction, the majority of Democrats believe that the world can keep using fossil fuels and still halt global heating anyway. In talking to these people, you will need to address their fears that renewable energy is too expensive or unreliable; gently trouble the complacent assumption that wealth will shield the Global North from climate devastation; wake people up to the fact that China is way ahead of the US in the race to net-zero; and disabuse them of the notion that carbon dioxide removal can enable the world to decarbonize ongoing fossil energy consumption. Of course, you don't want to use what you may have learned in this book to persuade anyone that they're wrong. If you make someone feel like they lost an argument about climate change they'll engage with climate change even less. You simply want to be prepared to help educate concerned Democrats and Independents even as you turn their attention to the people promoting fossil fuels and inspire them to oppose those people.

Having read this book, you are prepared to do this work. Indeed you are well armed! You have the beginnings of a new language of climate politics that can help get fossil fuels out of our lives, transform our systems, and bring global heating to an end. So go out and use your voice on behalf of the people you love most, knowing that by initiating conversations about the climate and countering fossil-fuel propaganda you're fighting to save not only their future, but the future of all the magnificent beings who dwell in our planet's life-giving embrace.

Acknowledgments

I would first like to thank my friends and colleagues who answered my questions about climate science, classical economics, systems thinking, resource extraction, plastics manufacturing, carbon pricing, integrated assessment modeling, Chinese policymaking, Congressional negotiations, backroom dealings at the COPs, and many more topics besides. I couldn't have navigated my way through my research without them. Thank you, Rebecca Altman, Kevin Anderson, Peter Brannen, Robert Brulle, Ana Unruh Cohen, Jonathan Coomey, Tan Copsey, Judith Enck, Sam Geall, Neil Grant, Emily Grubert, Zeke Hausfather, Ferris Jabr, Paul Kelleher, Alaa Al Khourdajie, Andreas Lichtenberger, Bob Litterman, Carroll Muffett, James Murray, Lauri Myllyvirta, Hector Pollitt, Thea Riofrancos, Elizabeth Sawin, June Sekera, Anat Shenker-Osorio, Jean Su, Nathan Tankus, Aaron Thierry, Michael Tobis, and Xiaoying You.

The Language of Climate Politics surely owes its existence to the people who so kindly helped me shape my book proposal. Thank you, Nicole Bond, Anthony Discenza, Eric Holthaus, Anya Kamenetz, Paul La Farge (you are missed), Michael Mann, Bill McKibben, Emily Raboteau, Elizabeth Rush, Margaret Klein Salamon, Michael Webber, and Ted Weinstein.

I am profoundly grateful to the extremely generous readers who took time out of their own projects to give me notes as I drafted my chapters. Their insightful suggestions hugely improved this book. Thank you, Jonathan Coomey, Meehan Crist, Bathsheba Demuth, Andrew Dessler, Pierre Friedlingstein, Ieva Jusionyte, Kate Mackenzie, Bill McKibben, Gregg Mitman, Ramez Naam, Edward Parson, Hector

Pollitt, Julia Steinberger, Gernot Wagner, David Wallace-Wells, and Travis Williams.

Special thanks goes to Zeke Hausfather, Glen Peters, and Hannah Ritchie for reading my critiques of their climate communication with remarkable graciousness and integrity.

I can't imagine having written this book without Michelle Niemann. Our weekly check-ins helped me maintain my sense of humor, figure out what I wanted to say next, and move forward with renewed resolve. Thomas LeBien provided swift editorial support when I needed it, thank goodness. David Kortava deftly fact-checked the entire manuscript and helped me update many of my citations, for which I am extremely grateful. (Any errors that may have crept in later are my own!)

One of my oldest friends, Susan Zieger, was writing a serious nonfiction book over the exact period that I was writing *The Language of Climate Politics*. It was a joy to share this process with her. I can only hope I offered her a fraction of the editorial wisdom and personal encouragement that she so graciously bestowed on me.

My book found its perfect home at Oxford University Press. I was so fortunate to work with my editor, the ever-astute Dave McBride; he really helped me chart a straight course between scholarly rigor and activist fervor. The rest of the team was tremendous, too. Sarah Ebel squired the manuscript through production with exquisite patience. Rachel Perkins designed a striking and witty book cover. (Thanks to Jacquelyn Gill and Mark Woloschuk for helping me clarify my own design ideas. And hat tip to Ed Hawkins for the cover's pun on his climate stripes!) Sheila Oakley did meticulous copyediting. Bridget M. Austiguy-Preschel served as a phenomenal resource during the typesetting phase. And Emily Tobin led a fantastic publicity campaign. I am deeply grateful to them all.

Immeasurable thanks go to Sarah Fuentes, my agent, who gave me wonderful referrals and answered all my questions as fast as I could ask them—and who at every phase of this project was a brilliant reader and a fierce advocate. Thank you, Sarah! Big thanks go also to Brian

Ulicky for coming on board with his formidable eloquence and invaluable savvy.

My greatest debt is to my *sine qua non*, my amazing husband Neal Cardwell. He read every word of this book, did everything possible to support it, and always had unshakeable confidence in my dreams. I don't know how I ever got so lucky.

The Language of Climate Politics is dedicated to our beautiful son, Teddy, who is the joy of my life. With all my heart, I hope this book helps create the future that he, and every child in the world, deserves, but no matter what happens I want him to know that his mother tried.

Notes

PREFACE

1. World Bank, *Poverty and Shared Prosperity 2022: Correcting Course* (The World Bank, 2022), 7, https://doi.org/10.1596/978-1-4648-1893-6.
2. Zach Christensen, "Economic Poverty Trends: Global, Regional and National" (Bristol, UK: Development Initiatives, February 2023), 8, https://devinit-prod-static.ams3.cdn.digitaloceanspaces.com/media/documents/Economic_poverty_factsheet_February_2023_JRJ2Y4f.pdf.
3. "Over the last 5,000 years, indigenous populations in [the Amazon] region coexisted with, and helped maintain, large expanses of relatively unmodified forest, as they continue to do today." See Dolores R. Piperno et al., "A 5,000-Year Vegetation and Fire History for *Tierra Firme* Forests in the Medio Putumayo-Algodón Watersheds, Northeastern Peru," *Proceedings of the National Academy of Sciences*, June 2021, 1, https://doi.org/10.1073/pnas.2022213118.

INTRODUCTION

1. Robert J. Brulle, "Institutionalizing Delay: Foundation Funding and the Creation of U.S. Climate Change Counter-Movement Organizations," *Climatic Change 122*, no. 4 (February 2014): 681–94, https://doi.org/10.1007/s10584-013-1018-7; Geoffrey Supran and Naomi Oreskes, "Assessing ExxonMobil's Climate Change Communications (1977–2014)," *Environmental Research Letters 12*, no. 8 (August 1, 2017): 084019, https://doi.org/10.1088/1748-9326/aa815f; Christian Downie, "Ad Hoc Coalitions in the U.S. Energy Sector: Case Studies in the Gas, Oil, and Coal Industries," *Business and Politics 20*, no. 4 (December 2018): 643–68, https://doi.org/10.1017/bap.2018.18; Robert J. Brulle, "Networks of Opposition: A Structural Analysis of U.S. Climate Change

Countermovement Coalitions 1989–2015," *Sociological Inquiry 91*, no. 3 (August 2021): 603–24, https://doi.org/10.1111/soin.12333.

2. Glenn Kessler, "Vivek Ramaswamy Says 'Hoax' Agenda Kills More People than Climate Change," *Washington Post*, August 25, 2023, https://www.washingtonpost.com/politics/2023/08/25/vivek-ramaswamy-says-hoax-agenda-kills-more-people-than-climate-change/.
3. Piers M. Forster et al., "Indicators of Global Climate Change 2022: Annual Update of Large-Scale Indicators of the State of the Climate System and Human Influence," *Earth System Science Data* 15, no. 6 (June 8, 2023): 2295–2327, https://doi.org/10.5194/essd-15-2295-2023.
4. UN Environment Programme, "Emissions Gap Report 2022" (United Nations, October 21, 2022), http://www.unep.org/resources/emissions-gap-report-2022.
5. "If today's energy infrastructure was to be operated until the end of the typical lifetime in a manner similar to the past, we estimate that this . . . is around 30% more than the remaining total CO_2 budget consistent with limiting global warming to 1.5 °C with a 50% probability." See International Energy Agency, "Net Zero by 2050—A Roadmap for the Global Energy Sector" (Paris, October 2021), 39, https://iea.blob.core.windows.net/assets/deebef5d-0c34-4539-9d0c-10b13d840027/NetZeroby2050-ARoadmapfortheGlobalEnergySector_CORR.pdf. "Projected cumulative future CO_2 emissions over the lifetime of existing fossil fuel infrastructure without additional abatement exceed the total cumulative net CO_2 emissions in pathways that limit warming to 1.5°C (>50%) with no or limited overshoot. They are approximately equal to total cumulative net CO_2 emissions in pathways that limit warming to 2°C with a likelihood of 83% . . . Limiting warming to 2°C (>67%) or lower will result in stranded assets. About 80% of coal, 50% of gas, and 30% of oil reserves cannot be burned and emitted if warming is limited to 2°C. Significantly more reserves are expected to remain unburned if warming is limited to 1.5°C. (high confidence)." See Intergovernmental Panel on Climate Change, "Climate Change 2023: Synthesis Report," 2023, 24, https://www.ipcc.ch/site/assets/uploads/2023/03/Doc5_Adopted_AR6_SYR_Longer_Report.pdf.
6. Josh Gabbatiss, "Analysis: Shell Admits 1.5C Climate Goal Means Immediate End to Fossil Fuel Growth," *Carbon Brief*, April 20, 2023, https://www.carbonbrief.org/analysis-shell-admits-1-5c-climate-goal-means-immediate-end-to-fossil-fuel-growth/.
7. Scott Waldman, "'Drill, Frack, Burn Coal': Republicans Echo Trump at Presidential Debate," *E&E News*, August 24, 2023, https://subscriber.

politicopro.com/article/eenews/2023/08/24/drill-frack-burn-coal-republicans-echo-trump-at-presidential-debate-cw-00112668.

8. Scott Waldman, "Conservatives Have Already Written a Climate Plan for Trump's Second Term," *E&E News*, July 26, 2023, https://subscriber.politicopro.com/article/eenews/2023/07/26/battle-plan-how-the-far-right-would-dismantle-climate-programs-00107498.
9. "Biden Administration Oil, Gas Drilling Approvals Outpace Trump's," Center for Biological Diversity, accessed August 31, 2023, https://biologicaldiversity.org/w/news/press-releases/biden-administration-oil-gas-drilling-approvals-outpace-trumps-2023-01-24.
10. Climate Action Tracker, New Climate Institute, and Climate Analytics, "Countdown to COP28: Time for the World to Focus on Oil and Gas Phase-out, Renewables Target – Not Distractions like CCS," 9, June 2023, https://climateactiontracker.org/documents/1144/CAT_2023-06-08_Briefing_PhaseOutOilGas.pdf.
11. U.S. Energy Information Administration (EIA), "EIA Expects U.S. Fossil Fuel Production to Reach New Highs in 2023," accessed August 31, 2023, https://www.eia.gov/todayinenergy/detail.php?id=50978.
12. U.S. Energy Information Administration (EIA), "Short-Term Energy Outlook," accessed January 19, 2024, https://www.eia.gov/outlooks/steo/.
13. Romain Ioualalen and Kelly Trout, "Planet Wreckers: How 20 Countries' Oil and Gas Extraction Plans Risk Locking in Climate Chaos" (Washington, D.C.: Oil Change International, September 2023), 17.
14. UN Environment Programme, "Emissions Gap Report 2022."
15. Zeke Hausfather and Glen P. Peters, "Emissions—the 'Business as Usual' Story Is Misleading," *Nature* 577, no. 7792 (January 30, 2020): 618–20, https://doi.org/10.1038/d41586-020-00177-3.
16. Intergovernmental Panel on Climate Change, "Sixth Assessment Report, Working Group II: Impacts, Adaptation, and Vulnerability. Overarching Frequently Asked Questions and Answers," June 2023, https://www.ipcc.ch/report/ar6/wg2/downloads/faqs/IPCC_AR6_WGII_Overaching_OutreachFAQ6.pdf.
17. Alec Tyson et al., "Americans Largely Favor U.S. Taking Steps to Become Carbon Neutral by 2050," *Pew Research Center Science & Society*, March 1, 2022, https://www.pewresearch.org/science/2022/03/01/americans-largely-favor-u-s-taking-steps-to-become-carbon-neutral-by-2050.
18. Alec Tyson et al., "What the Data Says about Americans' Views of Climate Change," *Pew Research Center*, accessed August 31, 2023, https://

www.pewresearch.org/short-reads/2023/08/09/what-the-data-says-about-americans-views-of-climate-change/. It's worth noting the generation gap here: Among Democrats and Independents who lean Democratic, 58% of those ages 18 to 29 favor phasing out fossil fuels entirely, compared with 42% of Democrats 65 and older.

CHAPTER 1

1. Peter Wallison and Benjamin Zycher, "This Winter We Will See the Dangerous Results of Climate Alarmism," *American Enterprise Institute (blog)*, accessed August 23, 2023, https://www.aei.org/articles/this-winter-we-will-see-the-dangerous-results-of-climate-alarmism/; for AEI funding sources, see InfluenceWatch, "American Enterprise Institute," accessed August 23, 2023, https://www.influencewatch.org/non-profit/american-enterprise-institute/.
2. H. Sterling Burnett, "Climate-Change Alarmists Are Getting More Delusional in Their Predictions," *The Heartland Institute (blog)*, accessed August 23, 2023, https://heartland.org/opinion/climate-change-alarmists-are-getting-more-delusional-in-their-predictions/.
3. "The Climate Cult Alarmists Are Waging a War on the American People" (Fox News Video), accessed August 23, 2023, https://www.foxnews.com/video/6309742772112.
4. Mia McCarthy, "Ron Johnson Says Climate Change Would Help Americans," *E&E News*, April 27, 2023, https://subscriber.politicopro.com/article/eenews/2023/04/27/ron-johnson-says-climate-change-would-help-americans-00094066.
5. Robert Ward, "Who Are 'Lukewarmers' and How Should Climate Change Researchers Respond to Them?," *American Geophysical Union, Fall Meeting*, 2018, https://ui.adsabs.harvard.edu/abs/2018AGUFMPA42C..10W.
6. Bjørn Lomborg, *False Alarm: How Climate Change Panic Costs Us Trillions, Hurts the Poor, and Fails to Fix the Planet* (New York: Basic Books, 2020), 6, 9.
7. See also Stefan Rahmstorf, "Bjørn Lomborg, just a scientist with a different opinion?," *Real Climate (blog)*, August 31, 2015, https://www.realclimate.org/index.php/archives/2015/08/bjorn-lomborg-just-a-scientist-with-a-different-opinion/; Bob Ward, "A Closer Examination of the Fantastical Numbers in Bjorn Lomborg's New Book," *Grantham Research Institute on Climate Change and the Environment (blog)*, accessed August 23, 2023, https://www.lse.ac.uk/granthaminstitute/news/a-clo

ser-examination-of-the-fantastical-numbers-in-bjorn-lomborgs-new-book/; Graham Readfearn, "The World Likely Just Had Its Hottest Month on Record. What a Time to Be a Climate Science Denier," *The Guardian*, August 3, 2023, https://www.theguardian.com/environment/commentisfree/2023/aug/03/the-world-likely-just-had-its-hottest-month-on-record-what-a-time-to-be-a-climate-science-denier.

8. Lomborg, *False Alarm*, 4.
9. YouGov, "International Poll: Most Expect to Feel Impact of Climate Change, Many Think It Will Make Us Extinct," September 15, 2019, https://yougov.co.uk/topics/politics/articles-reports/2019/09/15/international-poll-most-expect-feel-impact-climate.
10. Full disclosure: Lomborg has called me an "alarmist" on X, once known as Twitter. See https://twitter.com/BjornLomborg/status/1460539762968825857?s=20.
11. "Breakthrough Dialogue 2022: Progress Problems," The Breakthrough Institute, accessed August 23, 2023, https://thebreakthrough.org/events/bti-dialogue-progress-problems.
12. John Asafu-Adjaye et al., "An Ecomodernist Manifesto," *The Breakthrough Institute,* accessed August 23, 2023, http://www.ecomodernism.org.
13. J. Lelieveld et al., "Effects of Fossil Fuel and Total Anthropogenic Emission Removal on Public Health and Climate," *Proceedings of the National Academy of Sciences 116*, no. 15 (April 9, 2019): 7192–97, https://doi.org/10.1073/pnas.1819989116.
14. Asafu-Adjaye et al., "An Ecomodernist Manifesto."
15. Ted Nordhaus, "Marketing Catastrophism," *The Breakthrough Institute*, accessed August 25, 2023, https://thebreakthrough.org/journal/climate-change-banned-words/marketing-catastrophism. Full disclosure: Nordhaus has accused me of "catastrophism" (another word for "alarmism") in print. See Ted Nordhaus, "Am I the Mass Murderer?," The Breakthrough Institute, accessed August 25, 2023, https://thebreakthrough.org/journal/no-16-spring-2022/am-i-the-mass-murderer.
16. Joëlle Gergis, "A Climate Scientist's Take on Hope," in *Not Too Late: Changing the Climate Story from Despair to Possibility*, ed. Rebecca Solnit and Thelma Young Lutunatabua (Chicago: Haymarket Books, 2023), 40.
17. Hannah Ritchie, "Stop Telling Kids They'll Die from Climate Change," *WIRED UK*, accessed August 25, 2023, https://www.wired.co.uk/article/climate-crisis-doom.
18. The US-China Business Council, "China's Strategic Emerging Industries: Policy, Implementation, Challenges, and Recommendations" (Washington, D.C.: The US-China Business Council, March 2013), https://www.uschina.org/sites/default/files/sei-report.pdf.

19. According to the Yale Program on Climate Change Communication, the alarmed "strongly support climate policies" and are "most likely to engage in political activism." See Anthony Leiserowitz et al., "Global Warming's Six Americas" (New Haven, CT: Yale Program on Climate Change Communication and George Mason University Center for Climate Change Communication, 2015), https://climatecommunication.yale.edu/visualizations-data/six-americas/.
20. Hiroko Tabuchi, "101°F in the Ocean Off Florida: Was It a World Record?," *New York Times*, July 26, 2023, sec. Climate, https://www.nytimes.com/2023/07/26/climate/florida-100-degree-water.html.
21. António Guterres, "Secretary-General's Opening Remarks at Press Conference on Climate" (The United Nations, New York, NY), accessed August 25, 2023, https://www.un.org/sg/en/content/sg/speeches/2023-07-27/secretary-generals-opening-remarks-press-conference-climate.
22. Keynyn Brysse et al., "Climate Change Prediction: Erring on the Side of Least Drama?," *Global Environmental Change 23*, no. 1 (February 2013): 327–28, https://doi.org/10.1016/j.gloenvcha.2012.10.008.
23. Bob Watson, "2012 Frontiers of Geophysics Lecture," https://www.youtube.com/watch?v=YafoDGVAJAg. This clip was brought to my attention by Dr. Aaron Thierry.
24. *Climate Leaders: Chris Rapley CBE*, 2021, https://www.youtube.com/watch?v=JxKeacY8ULQ.
25. Quoted in Alastair McIntosh, "How to Be a Climate-Change Activist without Becoming an Alarmist," *Plough*, May 15, 2022, https://www.plough.com/en/topics/justice/environment/how-to-be-a-climate-change-activist.
26. Chris Colose [@CColose], "There's No Global Boiling. Please Don't Let That Be a Thing," *Tweet, Twitter*, July 27, 2023, https://twitter.com/CColose/status/1684674407585722368.
27. Chris Colose [@CColose], "Some People Have Criticized Me for Pushing Back on 'Global Boiling,' as If It Were an Unnecessary Fact Check. But the Facts Aren't the Issue. I Know Most People Don't Think the Oceans Will Boil (and Wouldn't until Temperatures in Excess of 600 K, Nearly Venus). It's Because These . . .," *Tweet, Twitter*, July 29, 2023, https://twitter.com/CColose/status/1685437262207635456.
28. Katharine Hayhoe [@KHayhoe], "@CColose 💯," *Tweet, Twitter*, July 30, 2023, https://twitter.com/KHayhoe/status/1685712030638989312.
29. See Katharine Hayhoe, *Saving Us: A Climate Scientist's Case for Hope and Healing in a Divided World* (New York: One Signal Publishers, 2021).

30. Michael E. Mann, *The New Climate War: The Fight to Take Back Our Planet* (New York: PublicAffairs, 2021), 182.
31. Saffron O'Neill and Sophie Nicholson-Cole, "'Fear Won't Do It': Promoting Positive Engagement with Climate Change through Visual and Iconic Representations," *Science Communication 30*, no. 3 (March 2009): 355–79, https://doi.org/10.1177/1075547008329201. Of course it makes sense that new mothers occupied by the demands of young children and worried about money would turn away from a problem represented as the suffering of bears at the North Pole, but not every subgroup will respond to images of climate change, or any climate message, in the same way. Indeed, two thousand years' scholarship on rhetoric has counseled speakers to tailor their messages to their audiences precisely because class, education, gender, partisanship, and so on lead different groups of people to hear information differently. See, e.g., Aristotle on the differences between the broad-strokes oratory addressed to public assemblies and the highly finished style appropriate for judges in courtroom (*Rhetoric*, Book III, Section 12) or Cicero on the difference between a subtle approach or a direct opening depending on whether the subject might alienate listeners or whether listeners have already been persuaded by the opposition (*Rhetorica ad Herennium*, Book I, Sections 4–7), or, for a contemporary example, the cognitive scientist and messaging strategist George Lakoff on the ways that effective political communication appeals to its audiences' partisan values (*Don't Think of an Elephant*). For more on fear messages in climate communication, see Genevieve Guenther, "Communicating the Climate Emergency: Imagination, Emotion, Action," in *Standing up for a Sustainable World: Voices of Change*, ed. Claude Henry, Johan Rockström, and N. H. Stern (Cheltenham, UK: Edward Elgar Publishing, 2020), 401–8.
32. See Mann, *The New Climate War*, 182–205; and "Surrender" in William F. Lamb et al., "Discourses of Climate Delay," *Global Sustainability 3* (2020): e17, https://doi.org/10.1017/sus.2020.13.
33. Seaver Wang et al., "Mechanisms and Impacts of Earth System Tipping Elements," *Reviews of Geophysics 61*, no. 1 (March 2023): e2021RG000757, https://doi.org/10.1029/2021RG000757.
34. I've taken my assumption for the global median value of coastal slope from Rafael Almar et al., "A Global Analysis of Extreme Coastal Water Levels with Implications for Potential Coastal Overtopping," *Nature Communications 12*, no. 1 (June 18, 2021): 3775, https://doi.org/10.1038/s41467-021-24008-9.

35. Timothy M. Lenton et al., "Climate Tipping Points—Too Risky to Bet Against," *Nature* 575, no. 7784 (November 2019): 592–95, https://doi.org/10.1038/d41586-019-03595-0.
36. Caroline Hickman et al., "Young People's Voices on Climate Anxiety, Government Betrayal and Moral Injury: A Global Phenomenon," *SSRN Electronic Journal*, 2021, https://doi.org/10.2139/ssrn.3918955.
37. One of the best of these, late philosopher Richard Rorty's *Achieving Our Country*, so perceptively analyzed the way that neoliberal thinking across the partisan divide leads to pervasive disengagement that it predicted the rise of Donald Trump.
38. Alexandria Ocasio-Cortez, "Alexandria Ocasio-Cortez on Instagram: 'It's Been a While since My Last Live! Just Saying Hi, Student Loan Updates, Talking about Climate Doomerism and Some Rich Guy in a Too-Tight Suit Trying to Fight Me on 5th Ave,'" *Instagram*, August 5, 2023, https://www.instagram.com/tv/CvitoiEoS2x/.
39. Roy Scranton, "No Happy Ending: On Bill McKibben's 'Falter' and David Wallace-Wells's 'The Uninhabitable Earth,'" *Los Angeles Review of Books*, June 3, 2019, https://lareviewofbooks.org/article/no-happy-ending-on-bill-mckibbens-falter-and-david-wallace-wellss-the-uninhabitable-earth/.
40. Zeke Hausfather, "Explainer: Will Global Warming 'Stop' as Soon as Net-Zero Emissions Are Reached?," *Carbon Brief*, April 29, 2021, https://www.carbonbrief.org/explainer-will-global-warming-stop-as-soon-as-net-zero-emissions-are-reached/.
41. Anthony Leiserowitz et al., "Climate Change in the American Mind: Beliefs and Attitudes, Spring 2023" (New Haven, CT: Yale Program on Climate Change Communication and George Mason University Center for Climate Change Communication, June 8, 2023), https://climatecommunication.yale.edu/publications/climate-change-in-the-american-mind-beliefs-attitudes-spring-2023/.
42. Anthony Leiserowitz et al., "Global Warming's Six Americas" (New Haven, CT: Yale Program on Climate Change Communication and George Mason University Center for Climate Change Communication, December 2022), https://climatecommunication.yale.edu/about/projects/global-warmings-six-americas/.
43. Stella Levantesi and Giulio Corsi, "Climate 'Realism' Is the New Climate Denial," *The New Republic*, August 6, 2020, https://newrepublic.com/article/158797/climate-change-alarmism-greta-thunberg-naomi-seibt.
44. Public Citizen, "Foxic: Fox News Network's Dangerous Climate Denial 2019," August 13, 2019, https://www.citizen.org/wp-content/uploads/public-citizen-fox-new-climate-denial-report-2019.pdf.

45. Helier Cheung, "What Does Trump Actually Believe on Climate Change?," *BBC News*, January 23, 2020, sec. US & Canada, https://www.bbc.com/news/world-us-canada-51213003.
46. Brad Plumer and Nadja Popovich, "Why Half a Degree of Global Warming Is a Big Deal," *New York Times*, sec. climate, October 7, 2018, https://www.nytimes.com/interactive/2018/10/07/climate/ipcc-report-half-degree.html.
47. António Guterres, "Secretary-General on Climate Action [as Delivered]" (The United Nations, New York, NY, May 30, 2017), https://www.un.org/sg/en/content/sg/statement/2017-05-30/secretary-general-climate-action-delivered.
48. The Editorial Board, "Wake Up, World Leaders. The Alarm Is Deafening," *New York Times*, October 9, 2018, sec. Opinion, https://www.nytimes.com/2018/10/09/opinion/climate-change-ipcc-report.html.
49. Alexis Berg et al., "Scientists Explain What New York Magazine Article on 'The Uninhabitable Earth' Gets Wrong," *Climate Feedback (blog)*, July 12, 2017, https://climatefeedback.org/evaluation/scientists-explain-what-new-york-magazine-article-on-the-uninhabitable-earth-gets-wrong-david-wallace-wells/.
50. David Wallace-Wells, "Beyond Catastrophe: A New Climate Reality Is Coming Into View," *New York Times*, October 26, 2022, sec. Magazine, https://www.nytimes.com/interactive/2022/10/26/magazine/climate-change-warming-world.html.
51. Thomas Harrisson, "Explainer: How 'Shared Socioeconomic Pathways' Explore Future Climate Change," *Carbon Brief*, April 19, 2018, https://www.carbonbrief.org/explainer-how-shared-socioeconomic-pathways-explore-future-climate-change/.
52. David Wallace-Wells, "Green New Deal Media," accessed August 25, 2023, https://www.gndmedia.co.uk/podcast-episodes/the-uninhabitable-earth-no-more-with-david-wallace-wells.
53. Greta Thunberg, "'Our House Is on Fire,'" *The Guardian*, January 25, 2019, sec. Environment, https://www.theguardian.com/environment/2019/jan/25/our-house-is-on-fire-greta-thunberg16-urges-leaders-to-act-on-climate.
54. Eliza Barclay, "How Big Was the Global Climate Strike? 4 Million People, Activists Estimate," *Vox*, September 20, 2019, https://www.vox.com/energy-and-environment/2019/9/20/20876143/climate-strike-2019-september-20-crowd-estimate; Matthew Taylor and John Bartlett, "Fresh Wave of Climate Strikes Takes Place around the World," *The Guardian*, September 27, 2019, sec. Environment, https://www.theguard

ian.com/environment/2019/sep/27/fresh-wave-of-climate-strikes-take-place-around-the-world.

55. Matthew Goldberg et al. "For the First Time, the Alarmed Are Now the Largest of Global Warming's Six Americas," *Yale Program on Climate Change Communication (blog),* accessed August 25, 2023, https://climatecommunication.yale.edu/publications/for-the-first-time-the-alarmed-are-now-the-largest-of-global-warmings-six-americas/.
56. Sharon Bernstein, "Trump, Biden Clash over Climate, Oil Industry in Final Debate," *Reuters*, October 23, 2020, sec. World News, https://www.reuters.com/article/uk-usa-election-debate-climate-change-idUKKBN2780HQ.
57. "Inflation Reduction Act of 2022," *Internal Revenue Service*, accessed August 25, 2023, https://www.irs.gov/inflation-reduction-act-of-2022.
58. UN Environment Programme, "Emissions Gap Report 2022," October 21, 2022, XVI, http://www.unep.org/resources/emissions-gap-report-2022.
59. Justin Ritchie and Hadi Dowlatabadi, "The 1000 GtC Coal Question: Are Cases of Vastly Expanded Future Coal Combustion Still Plausible?," *Energy Economics* 65 (June 1, 2017): 16–31, https://doi.org/10.1016/j.eneco.2017.04.015.
60. Zeke Hausfather and Justin Ritchie, "A 3C World Is Now 'Business as Usual,'" *The Breakthrough Institute (blog),* accessed August 25, 2023, https://thebreakthrough.org/issues/energy/3c-world. To be sure, Hausfather and Richie also said that 2.5°C would be the best the world could do without new policy (see p. 28).
61. Zeke Hausfather and Glen P. Peters, "Emissions—the 'Business as Usual' Story Is Misleading," *Nature* 577, no. 7792 (January 30, 2020): 618–20, https://doi.org/10.1038/d41586-020-00177-3.
62. Hausfather and Peters.
63. Roger Pielke and Justin Ritchie, "Distorting the View of Our Climate Future: The Misuse and Abuse of Climate Pathways and Scenarios," *Energy Research & Social Science* 72 (February 2021): 101890, https://doi.org/10.1016/j.erss.2020.101890.
64. Ross Douthat, "Neither Hot Nor Cold on Climate," *New York Times*, June 3, 2017, sec. Opinion, https://www.nytimes.com/2017/06/03/opinion/sunday/neither-hot-nor-cold-on-climate.html.
65. Ross Douthat, "The Contradictions of Climate Activism," *New York Times*, October 15, 2022, sec. Opinion, https://www.nytimes.com/2022/10/15/opinion/oil-energy-crisis-van-gogh.html.

66. Holman Jenkins Jr., "Climate Media vs. Climate Science," *Wall Street Journal*, April 13, 2021, https://www.wsj.com/articles/climate-media-vs-climate-science-11618355224.
67. David Wallace-Wells, "Here's Some Good News on Climate Change: Worst-Case Scenario Looks Unrealistic," *Intelligencer*, December 20, 2019, https://nymag.com/intelligencer/2019/12/climate-change-worst-case-scenario-now-looks-unrealistic.html.
68. Jenkins, "Climate Media vs. Climate Science."
69. Wallace-Wells, "Here's Some Good News on Climate Change."
70. Wallace-Wells, "Green New Deal Media."
71. David Wallace-Wells, "The New World: Envisioning Life After Climate Change," *New York Times*, October 26, 2022, sec. Magazine, https://www.nytimes.com/interactive/2022/10/26/magazine/visualization-climate-change-future.html.
72. Wallace-Wells, "Green New Deal Media."
73. Wallace-Wells, "Beyond Catastrophe."
74. Other journalists covered the revised warming estimates, although their articles tended to emphasize that warming of 3°C by 2100 would still be a catastrophe. See e.g. Robinson Meyer, "Are We Living Through Climate Change's Worst-Case Scenario?," *The Atlantic*, January 15, 2019, https://www.theatlantic.com/science/archive/2019/01/rcp-85-the-climate-change-disaster-scenario/579700/; Chris Mooney and Andrew Freedman, "We May Avoid the Very Worst Climate Scenario. But the Next-Worst Is Still Pretty Awful," *Washington Post*, January 30, 2020, https://www.washingtonpost.com/weather/2020/01/30/we-may-avoid-very-worst-climate-scenario-next-worst-is-still-pretty-awful/; Brad Plumer and Nadja Popovich, "Yes, There Has Been Progress on Climate. No, It's Not Nearly Enough," *New York Times*, October 25, 2021, sec. Climate, https://www.nytimes.com/interactive/2021/10/25/climate/world-climate-pledges-cop26.html; and Zahra Hirji, "'Bad For Humans': The World Is On Track to Warm 3 Degrees Celsius This Century," *BuzzFeed News*, October 30, 2021, https://www.buzzfeednews.com/article/zahrahirji/global-warming-3-degrees-celsius-impact.
75. Online Survey of 1000 Adults Nationwide August 21-24, 2020, End Climate Silence and Lake Research Partners.
76. Anthony Leiserowitz, "Global Warming's Six Americas, September 2021" (New Haven, CT: Yale Program on Climate Change Communication and George Mason University Center for Climate Change Communication, January 12, 2022), https://climatecommun

ication.yale.edu/publications/global-warmings-six-americas-september-2021/; Anthony Leiserowitz, "Global Warming's Six Americas, December 2022" (New Haven, CT: Yale Program on Climate Change Communication and George Mason University Center for Climate Change Communication, March 14, 2023), https://climatecommunication.yale.edu/publications/global-warmings-six-americas-december-2022/.

77. Intergovernmental Panel on Climate Change, "Climate Change 2023: Synthesis Report, Summary for Policymakers," 2023, 14, Figure SPM.3, https://www.ipcc.ch/report/ar6/syr/downloads/report/IPCC_AR6_SYR_SPM.pdf.
78. "Climate Change Could Be Worse than Our Worst-Case Analysis: Texas A&M's Dessler," *CNBC*, accessed August 25, 2023, https://www.cnbc.com/video/2021/07/21/climate-change-could-be-worse-than-our-worst-case-analysis-texas-ams-dessler.html.
79. Scott Dance, "A New Climate Reality: Less Warming, but Worse Impacts on the Planet," *Washington Post*, January 8, 2023, https://www.washingtonpost.com/climate-environment/2023/01/06/climate-change-scenarios-extremes/.
80. "Chief Coroner's Statement on Public Safety during High Temperatures," *BC Gov News,* July 30, 2021, https://news.gov.bc.ca/releases/2021PSSG0071-001523; "Death Toll from Record-Breaking Heat Wave Hits 116 in Oregon," *AP News*, July 7, 2021, https://apnews.com/article/oregon-heat-waves-2f6a8cafdb82f91792cff19fe10367e6; John Ryan, "2021 Heat Wave Is Now the Deadliest Weather-Related Event in Washington History," *KUOW*, July 20, 2021, https://www.kuow.org/stories/heat-wave-death-toll-in-washington-state-jumps-to-112-people.
81. Paul J. Schramm et al., "Heat-Related Emergency Department Visits during the Northwestern Heat Wave—United States, June 2021," *MMWR. Morbidity and Mortality Weekly Report 70*, no. 29 (July 23, 2021): 1020–21, https://doi.org/10.15585/mmwr.mm7029e1.
82. Samuel Bartusek, Kai Kornhuber, and Mingfang Ting, "2021 North American Heatwave Amplified by Climate Change-Driven Nonlinear Interactions," *Nature Climate Change 12*, no. 12 (December 2022): 1143–50, https://doi.org/10.1038/s41558-022-01520-4.
83. Emily Bercos-Hickey et al., "Anthropogenic Contributions to the 2021 Pacific Northwest Heatwave," *Geophysical Research Letters 49*, no. 23 (December 16, 2022): e2022GL099396, https://doi.org/10.1029/2022GL099396.

84. Hugo Francisco de Souza, "Impact of the 2022 European Heat Wave: Over 60,000 Deaths Recorded," *News-Medical.Net,* accessed August 25, 2023, https://www.news-medical.net/news/20230717/Impact-of-the-2022-European-heat-wave-Over-60000-deaths-recorded.aspx; Jonathan Watts, "Record Heatwave 'Made Much More Likely' by Human Impact on Climate," *The Guardian*, August 2, 2019, sec. Environment, https://www.theguardian.com/environment/2019/aug/02/record-heatwave-made-much-more-likely-by-human-impact-on-climate.
85. "Extreme Heat in North America, Europe and China in July 2023 Made Much More Likely by Climate Change," *World Weather Attribution*, July 25, 2023, https://www.worldweatherattribution.org/extreme-heat-in-north-america-europe-and-china-in-july-2023-made-much-more-likely-by-climate-change/; Phil Helsel, "Phoenix Ends Record 31-Day Streak of above 110-Degree Weather," *NBC News*, August 1, 2023, https://www.nbcnews.com/news/us-news/phoenix-ends-record-31-day-streak-110-degree-weather-rcna97458; Jen Christensen, "It's So Hot in Arizona, Doctors Are Treating a Spike of Patients Who Were Burned by Falling on the Ground," *CNN*, July 24, 2023, https://www.cnn.com/2023/07/24/health/arizona-heat-burns-er/index.html.
86. Ian Livingston, "It's Midwinter, but It's over 100 Degrees in South America," *Washington Post*, August 2, 2023, https://www.washingtonpost.com/weather/2023/08/02/southamerica-record-winter-heat-argentina-chile/.
87. Aliza Chasan, "Unprecedented Ocean Temperatures 'Much Higher than Anything the Models Predicted,' Climate Experts Warn," *CBS News*, July 10, 2023, https://www.cbsnews.com/news/ocean-temperatures-higher-models-predicted-climate-experts-warn/.
88. Graham Readfearn, "'Something Weird Is Going On': Search for Answers as Antarctic Sea Ice Stays at Historic Lows," *The Guardian*, July 29, 2023, https://www.theguardian.com/world/2023/jul/29/something-weird-is-going-on-search-for-answers-as-antarctic-sea-ice-stays-at-historic-lows.
89. Ian Livingston, "Canada's Astonishing and Record Fire Season Finally Slows Down," *Washington Post*, October 18, 2023, https://www.washingtonpost.com/weather/2023/10/18/canada-historic-2023-wildfire-season-end/.
90. Sarah True, "What to Know about the Lethal Strain of Malaria Contracted in Maryland," *The Baltimore Banner*, August 23, 2023, https://www.thebaltimorebanner.com/community/public-health/malaria-in-maryl

and-3NEWOELD5ZHHVB37NFD4XZB5YE/; Christopher Walsh, "Flesh-Eating Bacteria a Threat in Warm Water," *The East Hampton Star*, accessed August 25, 2023, https://www.easthamptonstar.com/villages-health/2023824/flesh-eating-bacteria-threat-warm-water.

91. Gina Martinez, "Hilary Drenches Southern California with Record-Breaking Rainfall as Storm Wreaks Havoc," *CBC News*, August 21, 2023, https://www.cbsnews.com/news/tropical-storm-hilary-southern-california-nevada-rain-flooding/.
92. Alan Buis, "Could a Hurricane Ever Strike Southern California?," *NASA Climate Change: Vital Signs of the Planet*, October 6, 2012, https://climate.nasa.gov/news/802/could-a-hurricane-ever-strike-southern-california.
93. Roger Harrabin, "What Frightens Me about the Climate Crisis Is We Don't Know How Bad Things Really Are," *The Guardian*, July 25, 2023, sec. Opinion, https://www.theguardian.com/commentisfree/2023/jul/25/frightens-climate-crisis-do-not-know-how-bad-wildfires-greece.
94. Sam Meredith, "A Multitrillion-Dollar Carbon Bubble? Climate Chief Warns World Leaders over Fossil Fuel Plans," *CNBC*, August 16, 2023, https://www.cnbc.com/2023/08/16/climate-chief-world-leaders-at-risk-of-ignoring-fossil-fuel-time-bomb.html.
95. Roger Harrabin, "Faster Pace of Climate Change Is 'Scary', Former Chief Scientist Says," *BBC News*, September 15, 2019, sec. Science & Environment, https://www.bbc.com/news/science-environment-49689018.
96. Euan Nisbet, "Rising Methane Could Be a Sign That Earth's Climate Is Part-Way through a 'Termination-Level Transition,'" *The Conversation*, August 14, 2023, http://theconversation.com/rising-methane-could-be-a-sign-that-earths-climate-is-part-way-through-a-termination-level-transition-211211.
97. Stefan Rahmstorf, "What Is Happening in the Atlantic Ocean to the AMOC?," *RealClimate (blog)*, July 24, 2023, https://www.realclimate.org/index.php/archives/2023/07/what-is-happening-in-the-atlantic-ocean-to-the-amoc/.
98. Rahmstorf.
99. T. M. Lenton et al., "Tipping Elements in the Earth's Climate System," *Proceedings of the National Academy of Sciences 105*, no. 6 (February 12, 2008): 1786–93, https://doi.org/10.1073/pnas.0705414105.
100. Intergovernmental Panel on Climate Change, "Climate Change 2013: The Physical Science Basis, Summary for Policymakers," 2013, 22, https://

www.ipcc.ch/site/assets/uploads/2018/03/WGIAR5_SPM_brochure_en.pdf, emphasis in the original.

101. Intergovernmental Panel on Climate Change, "Climate Change 2021: The Physical Science Basis, Summary for Policymakers," 2021, 27,https://www.ipcc.ch/report/ar6/wg1/downloads/report/IPCC_AR6_WGI_SPM.pdf.
102. Niklas Boers, "Observation-Based Early-Warning Signals for a Collapse of the Atlantic Meridional Overturning Circulation," *Nature Climate Change 11*, no. 8 (August 2021): 680–88, https://doi.org/10.1038/s41558-021-01097-4; Simon L. L. Michel et al., "Early Warning Signal for a Tipping Point Suggested by a Millennial Atlantic Multidecadal Variability Reconstruction," *Nature Communications 13*, no. 1 (September 2, 2022): 5176, https://doi.org/10.1038/s41467-022-32704-3; Peter Ditlevsen and Susanne Ditlevsen, "Warning of a Forthcoming Collapse of the Atlantic Meridional Overturning Circulation," *Nature Communications 14*, no. 1 (July 25, 2023): 4254, https://doi.org/10.1038/s41467-023-39810-w.
103. Intergovernmental Panel on Climate Change, "Climate Change 2021: The Physical Science Basis, Summary for Policymakers," 27.
104. Peter Ditlevsen and Susanne Ditlevsen, "Atlantic Collapse: Q&A with Scientists behind Controversial Study Predicting a Colder Europe," *The Conversation*, August 10, 2023, http://theconversation.com/atlantic-collapse-qanda-with-scientists-behind-controversial-study-predicting-a-colder-europe-211221.
105. Intergovernmental Panel on Climate Change, "Climate Change 2021: The Physical Science Basis, Summary for Policymakers," 27.
106. "Expert Reaction to Paper Warning of a Collapse of the Atlantic Meridional Overturning Circulation," Science Media Centre, accessed August 25, 2023, https://www.sciencemediacentre.org/expert-reaction-to-paper-warning-of-a-collapse-of-the-atlantic-meridional-overturning-circulation/.
107. Ditlevsen and Ditlevsen, "Atlantic Collapse."
108. "Expert Reaction to Paper Warning of a Collapse of the Atlantic Meridional Overturning Circulation."
109. Thanks to geologically distinct interactions between the atmosphere and the ice sheets, this heating was caused by carbon dioxide concentrations of 400ppm. Those concentrations passed 420ppm in 2023. See Michael E. Mann, *Our Fragile Moment: How Lessons from Earth's Past Can Help Us Survive the Climate Crisis* (New York: PublicAffairs, 2023), 153–59.

110. Peter Brannen, "The Terrifying Warning Lurking in the Earth's Ancient Rock Record," *The Atlantic*, February 3, 2021, https://www.theatlantic.com/magazine/archive/2021/03/extreme-climate-change-history/617793/.
111. Richard E. Zeebe, Andy Ridgwell, and James C. Zachos, "Anthropogenic Carbon Release Rate Unprecedented during the Past 66 Million Years," *Nature Geoscience* 9, no. 4 (April 2016): 325–29, https://doi.org/10.1038/ngeo2681.
112. Personal Correspondence, July 11, 2023.
113. Sarah Kaplan, "Floods, Fires and Deadly Heat Are the Alarm Bells of a Planet on the Brink," *Washington Post*, July 14, 2023, https://www.washingtonpost.com/climate-environment/2023/07/12/climate-change-flooding-heat-wave-continue/.
114. Intergovernmental Panel on Climate Change, "Sixth Assessment Report, Working Group II: Impacts, Adaptation, and Vulnerability. Overarching Frequently Asked Questions and Answers," June 2023, https://www.ipcc.ch/report/ar6/wg2/downloads/faqs/IPCC_AR6_WGII_Overaching_OutreachFAQ6.pdf.
115. Thea Gregersen, Gisle Andersen, and Endre Tvinnereim, "The Strength and Content of Climate Anger," *Global Environmental Change 82* (September 2023): 102738, https://doi.org/10.1016/j.gloenvcha.2023.102738; see also, James M. Jasper, *The Emotions of Protest* (Chicago: The University of Chicago Press, 2018).
116. Erica Chenoweth and Maria J. Stephan, *Why Civil Resistance Works: The Strategic Logic of Nonviolent Conflict*, Columbia Studies in Terrorism and Irregular Warfare (New York: Columbia University Press, 2011).
117. Alec Tyson, Cary Funk, and Brian Kennedy, "What the Data Says about Americans' Views of Climate Change" (Washington, D.C.: Pew Research Center, August 9, 2023), https://www.pewresearch.org/short-reads/2023/08/09/what-the-data-says-about-americans-views-of-climate-change/.

CHAPTER 2

1. "Press Briefing by Ari Fleischer," March 28, 2001, https://georgewbush-whitehouse.archives.gov/news/briefings/20010328.html#KyotoTreaty.
2. Annie Lowrey, "The U.S. Is on the Path to Destruction," *The Atlantic*, September 18, 2020, https://www.theatlantic.com/ideas/archive/2020/09/4-degrees-celsius-election/616393/.

3. "News Release: A Return to the Paris Climate Agreement Will Raise Energy Costs and It Won't Solve Climate Change—United States Senator John Barrasso," January 20, 2021, https://www.barrasso.senate.gov/public/index.cfm/news-releases?ID=49154DB7-BC8F-402E-8902-D4A8AF826E84; "Press Release: Biden Chooses Paris over Pittsburgh | Ted Cruz | U.S. Senator for Texas," January 20, 2021, https://www.cruz.senate.gov/?p=press_release&id=5557.
4. Allison Fisher, "Fox's 'War on Earth Day,'" *Media Matters for America*, accessed April 27, 2021, https://www.mediamatters.org/fox-news/foxs-war-earth-day.
5. Online Survey of 1000 Adults Nationwide August 21–24, 2020, End Climate Silence and Lake Research Partners.
6. William Nordhaus, "An Optimal Transition Path for Controlling Greenhouse Gases," *Science 258*, no. 5086 (November 20, 1992): 1315–19, https://doi.org/10.1126/science.258.5086.1315.
7. William Nordhaus, *The Climate Casino: Risk, Uncertainty, and Economics for a Warming World* (New Haven, CT: Yale University Press, 2013), 191.
8. William Nordhaus, "A Review of the *Stern Review on the Economics of Climate Change*," *Journal of Economic Literature* 45, no. 3 (July 1, 2007): 692, https://doi.org/10.1257/jel.45.3.686.
9. Marlowe Hood, "Climate Economics Nobel May Do More Harm Than Good," *International Business Times*, July 6, 2020, https://www.ibtimes.com/climate-economics-nobel-may-do-more-harm-good-3006142?fbclid=IwAR0l3eHsrStS6JOfxDRottylA-2QrzCrobDR6IiDx681_7b_AJhpwua3ZmA.
10. Interagency Working Group on Social Cost of Greenhouse Gases, United States Government, "Technical Support Document: Social Cost of Carbon, Methane, and Nitrous Oxide Interim Estimates under Executive Order 13990" (Washington, D.C., February 2021), 22, https://www.whitehouse.gov/wp-content/uploads/2021/02/TechnicalSupportDocument_SocialCostofCarbonMethaneNitrousOxide.pdf.
11. William Nordhaus, "Revisiting the Social Cost of Carbon," *Proceedings of the National Academy of Sciences 114*, no. 7 (February 14, 2017): 1518–23, https://doi.org/10.1073/pnas.1609244114.
12. William Nordhaus and Joseph Boyer, "Requiem for Kyoto: An Economic Analysis of the Kyoto Protocol," *The Energy Journal 20*, no. 1 (September 1, 1999), https://doi.org/10.5547/ISSN0195-6574-EJ-Vol20-NoSI-5.
13. Nordhaus, *The Climate Casino*, 7. See also 176 and passim.

14. William Nordhaus, "Climate Change: The Ultimate Challenge for Economics," *American Economic Review 109*, no. 6 (June 1, 2019): 1991–2014, 451, https://doi.org/10.1257/aer.109.6.1991.
15. William Nordhaus, "Climate Change: The Ultimate Challenge for Economics," NobelPrize.org, 2018, https://www.nobelprize.org/uploads/2018/10/nordhaus-lecture.pdf .
16. Zeke Hausfather and Glen P. Peters, "Emissions–the 'Business as Usual' Story Is Misleading," *Nature 577*, no. 7792 (January 30, 2020): 619, https://doi.org/10.1038/d41586-020-00177-3.
17. Hood, "Climate Economics Nobel May Do More Harm Than Good"; see also Delavane Diaz and Frances Moore, "Quantifying the Economic Risks of Climate Change," *Nature Climate Change 7*, no. 11 (November 2017): 774–82, 774, https://doi.org/10.1038/nclimate3411: "A clear disconnect exists between the current scientific literature on climate change impacts, adaptation, and vulnerability (IAV) and the representation of climate-change impacts in these cost-benefit models."
18. Hood, "Climate Economics Nobel May Do More Harm Than Good."
19. Nordhaus, "Climate Change," 1992.
20. Nordhaus, *The Climate Casino*, 189.
21. Nordhaus, "A Review of the *Stern Review on the Economics of Climate Change*," 693.
22. See Gernot Wagner and Martin L. Weitzman, *Climate Shock: The Economic Consequences of a Hotter Planet* (Princeton, NJ: Princeton University Press, 2015), 63–65.
23. Richard L. Revesz et al., "Global Warming: Improve Economic Models of Climate Change," *Nature 508*, no. 7495 (April 2014): 174, https://doi.org/10.1038/508173a.
24. Steve Keen, "The Appallingly Bad Neoclassical Economics of Climate Change," *Globalizations*, September 1, 2020, 4–5, https://doi.org/10.1080/14747731.2020.1807856.
25. Revesz et al., "Global Warming," 175.
26. Noah Kaufman, "The Social Cost of Carbon in Taxes and Subsidies" (*Center on Global Energy Policy, Columbia University,* March 2018); Timothy M. Lenton et al., "Climate Tipping Points — Too Risky to Bet Against," *Nature 575*, no. 7784 (November 2019): 592–95, https://doi.org/10.1038/d41586-019-03595-0.
27. Kaufman, "The Social Cost of Carbon in Taxes and Subsidies," 15.
28. Tamma A. Carleton et al., "NBER Working Papers Series: Valuing the Global Mortality Consequences of Climate Change Accounting for Adaptation Costs and Benefits," n.d., 5–6; Peter H. Howard and

Thomas Sterner, "Few and Not So Far Between: A Meta-Analysis of Climate Damage Estimates," *Environmental and Resource Economics 68*, no. 1 (September 2017): 197–225, https://doi.org/10.1007/s10640-017-0166-z.

29. William Nordhaus, "Expert Opinion on Climatic Change," *American Scientist 82*, no. 1 (1994): 45–51, emphasis mine; see also Keen, "The Appallingly Bad Neoclassical Economics of Climate Change."
30. Thomas Stoerk, Gernot Wagner, and Robert E. T. Ward, "Policy Brief—Recommendations for Improving the Treatment of Risk and Uncertainty in Economic Estimates of Climate Impacts in the Sixth Intergovernmental Panel on Climate Change Assessment Report," *Review of Environmental Economics and Policy 12*, no. 2 (July 1, 2018): 374, https://doi.org/10.1093/reep/rey005.
31. Nordhaus, *The Climate Casino*, 144–45.
32. Nordhaus, "Climate Change," 2000.
33. Lint Barrage and William Nordhaus, "Policies, Projections, and the Social Cost of Carbon: Results from the DICE-2023 Model," *National Bureau of Economic Research, NBER Working Paper Series, Working Paper 31112* (April 2023), http://www.nber.org/papers/w31112.
34. Simon Dietz and Nicholas Stern, "Endogenous Growth, Convexity of Damage and Climate Risk: How Nordhaus' Framework Supports Deep Cuts in Carbon Emissions," *The Economic Journal 125*, no. 583 (March 1, 2015): 3, https://doi.org/10.1111/ecoj.12188.
35. Robert S. Pindyck, "The Use and Misuse of Models for Climate Policy," *Review of Environmental Economics and Policy 11*, no. 1 (January 1, 2017): 100, https://doi.org/10.1093/reep/rew012.
36. Diaz and Moore, "Quantifying the Economic Risks of Climate Change."
37. Howard and Sterner, "Few and Not So Far Between," 199.
38. Stoerk, Wagner, and Ward, "Policy Brief—Recommendations for Improving the Treatment of Risk and Uncertainty in Economic Estimates of Climate Impacts in the Sixth Intergovernmental Panel on Climate Change Assessment Report," 374.
39. Joseph Stiglitz, "The Economic Case for a Green New Deal," in *Winning the Green New Deal: Why We Must, How We Can*, ed. Varshini Prakash and Guido Girgenti, First Simon & Schuster trade paperback edition (New York: Simon & Schuster, 2020), 96.
40. Cristian Proistosescu and Gernot Wagner, "Uncertainties in Climate and Weather Extremes Increase the Cost of Carbon," *One Earth 2*, no. 6 (June 2020): 516, https://doi.org/10.1016/j.oneear.2020.06.002; David

J. Frame et al., "The Economic Costs of Hurricane Harvey Attributable to Climate Change," *Climatic Change 160*, no. 2 (May 2020): 271–81, https://doi.org/10.1007/s10584-020-02692-8.

41. Colin J. Carlson et al., "Climate Change Increases Cross-Species Viral Transmission Risk," *Nature 607*, no. 7919 (July 21, 2022): 555–62, https://doi.org/10.1038/s41586-022-04788-w.
42. Proistosescu and Wagner, "Uncertainties in Climate and Weather Extremes Increase the Cost of Carbon," 516.
43. Eric Maskin, "Toast to Marty Weitzman (Prepared for Marty's Retirement Dinner, October 11, 2018)," https://economics.harvard.edu/files/economics/files/toast_to_marty_weitzman_10.11.2018_e._maskin.pdf.
44. Rebecca Araten, "Harvard Economist Martin Weitzman, Known for Climate Change Scholarship, Dies at 77," *The Harvard Crimson*, September 6, 2019, https://www.thecrimson.com/article/2019/9/6/martin-weitzman-obituary/.
45. Martin L. Weitzman, "On Modeling and Interpreting the Economics of Catastrophic Climate Change," *Review of Economics and Statistics 91*, no. 1 (February 2009): 1, https://doi.org/10.1162/rest.91.1.1. Weitzman is quoting Richard Posner here.
46. Weitzman, 2.
47. Intergovernmental Panel on Climate Change, "AR5 Synthesis Report: Climate Change 2014," 2014, 223, https://www.ipcc.ch/report/ar5/syr/.
48. Wagner and Weitzman, Climate Shock, 55.
49. Wagner and Weitzman, 78–79.
50. Weitzman, "On Modeling and Interpreting the Economics of Catastrophic Climate Change," 5.
51. Martin L. Weitzman, "Fat Tails and the Social Cost of Carbon," *American Economic Review 104*, no. 5 (May 1, 2014): 544–46, https://doi.org/10.1257/aer.104.5.544.
52. Wagner and Weitzman, *Climate Shock*, 73; see also Martin L. Weitzman, "GHG Targets as Insurance against Catastrophic Climate Damages," *Journal of Public Economic Theory 14*, no. 2 (March 2012): 221–44, https://doi.org/10.1111/j.1467-9779.2011.01539.x.
53. Wagner and Weitzman, *Climate Shock*, 73; Kent D. Daniel, Robert B. Litterman, and Gernot Wagner, "Declining CO_2 Price Paths," *Proceedings of the National Academy of Sciences 116*, no. 42 (October 15, 2019): 20886–91, https://doi.org/10.1073/pnas.1817444116.

54. Weitzman, "On Modeling and Interpreting the Economics of Catastrophic Climate Change," 2.
55. Weitzman, 17.
56. Wagner and Weitzman, *Climate Shock*, 82.
57. Weitzman, "On Modeling and Interpreting the Economics of Catastrophic Climate Change," 18.
58. William Nordhaus, "Cowles Foundation Discussion Paper No. 1686: An Analysis of the Dismal Theorem" (*Cowles Foundation for Research in Economics, Yale University*, January 2009), 22.
59. Nordhaus, "A Review of the *Stern Review on the Economics of Climate Change*," 693.
60. Intergovernmental Panel on Climate Change, "Climate Change 2023: Synthesis Report, Summary for Policymakers," 2023, 14, Figure SPM.3, https://www.ipcc.ch/report/ar6/syr/downloads/report/IPCC_AR6_SYR_SPM.pdf.
61. Eduardo Porter, "Climate Deal Badly Needs a Big Stick," *New York Times*, June 2, 2015, sec. Business, https://www.nytimes.com/2015/06/03/business/energy-environment/climate-deal-badly-needs-a-big-stick.html.
62. Eric Roston, "The Man Who Got Economists to Take Climate Nightmares Seriously," *Bloomberg*, August 29, 2019, https://www.bloomberg.com/news/articles/2019-08-29/the-man-who-got-economists-to-take-climate-nightmares-seriously?sref=UIYAtP2Y.
63. Committee on Assessing Approaches to Updating the Social Cost of Carbon et al., *Valuing Climate Changes: Updating Estimation of the Social Cost of Carbon Dioxide* (Washington, D.C.: National Academies Press, 2017), https://doi.org/10.17226/24651.
64. Marshall Burke, Solomon M. Hsiang, and Edward Miguel, "Global Non-Linear Effect of Temperature on Economic Production," *Nature* 527, no. 7577 (November 2015): 238, 239, https://doi.org/10.1038/nature15725.
65. Marshall Burke, W. Matthew Davis, and Noah S. Diffenbaugh, "Large Potential Reduction in Economic Damages under UN Mitigation Targets," *Nature* 557, no. 7706 (May 2018): 549–53, https://doi.org/10.1038/s41586-018-0071-9.
66. Swiss Re, "Press Release: World Economy Set to Lose up to 18% GDP from Climate Change If No Action Taken, Reveals Swiss Re Institute's Stress-Test Analysis," April 22, 2021, https://www.swissre.com/media/press-release/nr-20210422-economics-of-climate-change-risks.html.

67. Yi-Ming Wei et al., "Self-Preservation Strategy for Approaching Global Warming Targets in the Post-Paris Agreement Era," *Nature Communications 11*, no. 1 (December 2020): 1624, https://doi.org/10.1038/s41467-020-15453-z.
68. Burke, Davis, and Diffenbaugh, "Large Potential Reduction in Economic Damages under UN Mitigation Targets," 4.
69. Martin C. Hänsel et al., "Climate Economics Support for the UN Climate Targets," *Nature Climate Change 10*, no. 8 (August 2020): 781–89, https://doi.org/10.1038/s41558-020-0833-x; Nicole Glanemann, Sven N. Willner, and Anders Levermann, "Paris Climate Agreement Passes the Cost-Benefit Test," *Nature Communications 11*, no. 1 (December 2020): 110, https://doi.org/10.1038/s41467-019-13961-1.
70. Bernardo A. Bastien-Olvera and Frances C. Moore, "Use and Non-Use Value of Nature and the Social Cost of Carbon," *Nature Sustainability*, September 28, 2020, https://doi.org/10.1038/s41893-020-00615-0.
71. Benjamin Zycher, "The Climate Left Attacks Nobel Laureate William D. Nordhaus" (*American Enterprise Institute,* July 2020), 2.
72. Eric Larson et al., "Net-Zero America: Potential Pathways, Infrastructure, and Impacts" (Princeton, NJ: Princeton University, December 15, 2020).
73. James H. Williams et al., "Carbon-Neutral Pathways for the United States," *AGU Advances 2*, no. 1 (March 2021), https://doi.org/10.1029/2020AV000284.
74. Marina Andrijevic et al., "COVID-19 Recovery Funds Dwarf Clean Energy Investment Needs," *Science 370*, no. 6514 (October 16, 2020): 298–300.
75. Nathaniel Bullard, "This Is the Dawning of the Age of the Battery," *Bloomberg.Com*, December 17, 2020, https://www.bloomberg.com/news/articles/2020-12-17/this-is-the-dawning-of-the-age-of-the-battery; BloombergNEF, "Battery Pack Prices Cited Below $100/kWh for the First Time in 2020, While Market Average Sits at $137/kWh," *BloombergNEF (blog),* December 16, 2020, https://about.bnef.com/blog/battery-pack-prices-cited-below-100-kwh-for-the-first-time-in-2020-while-market-average-sits-at-137-kwh/.
76. U.S. Energy Information Administration (EIA), "Utility-Scale Battery Storage Costs Decreased Nearly 70% between 2015 and 2018," accessed April 27, 2021, https://www.eia.gov/todayinenergy/detail.php?id=45596.
77. International Renewable Energy Agency (IRENA), "Renewable Power Generation Costs in 2019" (Abu Dhabi, United Arab Emirates:

International Renewable Energy Agency, 2020), https://www.irena.org/-/media/Files/IRENA/Agency/Publication/2020/Jun/IRENA_Power_Generation_Costs_2019.pdf?rev=77ebbae10ca34ef98909a59e39470906.

78. "Why Did Renewables Become So Cheap So Fast? And What Can We Do to Use This Global Opportunity for Green Growth?," *Our World in Data,* accessed April 27, 2021, https://ourworldindata.org/cheap-renewables-growth.
79. Dana Nuccitelli, "Fighting Climate Change: Cheaper than 'Business as Usual' and Better for the Economy," *Yale Climate Connections*, November 30, 2020, https://yaleclimateconnections-newspack.newspackstaging.com/2020/11/fighting-climate-change-cheaper-than-business-as-usual-and-better-for-the-economy/.
80. International Energy Agency, "Net Zero Roadmap: A Global Pathway to Keep the 1.5 °C Goal in Reach" (Paris, September 2023), 174, https://iea.blob.core.windows.net/assets/13dab083-08c3-4dfd-a887-42a3ebe533bc/NetZeroRoadmap_AGlobalPathwaytoKeepthe1.5CGoalinReach-2023Update.pdf.
81. Brian Kennedy, Cary Funk, and Alec Tyson, "What Americans Think about an Energy Transition from Fossil Fuels to Renewables" (Washington, D.C.: Pew Research Center, June 28, 2023), https://www.pewresearch.org/science/2023/06/28/what-americans-think-about-an-energy-transition-from-fossil-fuels-to-renewables/.
82. Amol Phadke et al., "2035: The Report" (Berkeley, CA: Goldman School of Public Policy, UC Berkeley, June 2020).
83. Coral Davenport, Lisa Friedman, and Jim Tankersley, "Biden's Bet on a Climate Transition Carries Big Risks," *New York Times*, April 24, 2021, sec. Business, https://www.nytimes.com/2021/04/24/business/bidens-climate-change.html.
84. Saul Griffith, Sam Calisch, and Laura Fraser, "Rewiring America: A Handbook for Winning the Climate Fight," July 2020, https://www.rewiringamerica.org/policy/rewiring-america-handbook.
85. Miranda Willson, "Energy Transitions: DOE Study: Net-Zero Emissions Feasible at Low Cost," February 2, 2021, https://www.eenews.net/stories/1063724093.
86. Drew Shindell et al., "Quantified, Localized Health Benefits of Accelerated Carbon Dioxide Emissions Reductions," *Nature Climate Change 8*, no. 4 (April 2018): 291–95, https://doi.org/10.1038/s41558-018-0108-y.

87. See Amol Phadke et al., "Illustrative Pathways to 100 percent Zero Carbon Power by 2035 without Increasing Customer Costs" (*Energy Innovation,* September 2020).
88. Robbie Orvis, "A 1.5 Celsius Pathway to Climate Leadership for the United States" (San Francisco, CA: Energy Innovation, February 2021); Phadke et al., "Illustrative Pathways"; Griffith, Calisch, and Fraser, "Rewiring America: A Handbook for Winning the Climate Fight"; Mark Paul, Anders Fremstad, and J.W. Mason, "Decarbonizing the US Economy: Pathways toward a Green New Deal" (*Roosevelt Institute,* June 2019); see also Matto Mildenberger and Leah Stokes, "The Trouble with Carbon Pricing," *Boston Review*, September 23, 2020, http://bostonreview.net/science-nature-politics/matto-mildenberger-leah-c-stokes-trouble-carbon-pricing.
89. Stiglitz, "The Economic Case for a Green New Deal," 104.
90. "Unemployment Rate by Educational Attainment and Sex," U.S. Department of Labor, accessed October 19, 2023, http://www.dol.gov/agencies/wb/data/latest-annual-data/working-women/Unemployment-Rate-by-Educational-Attainment-Sex.
91. Stephanie Kelton, *The Deficit Myth: Modern Monetary Theory and the Birth of the People's Economy* (PublicAffairs, 2021).
92. Ray Galvin and Noel Healy, "The Green New Deal in the United States: What It Is and How to Pay for It," *Energy Research & Social Science 67* (September 2020): 101529, https://doi.org/10.1016/j.erss.2020.101529; Yeva Nersisyan and L. Randall Wray, "How to Pay for the Green New Deal" (Poughkeepsie, NY: Levy Economics Institute, Bard College, 2019), https://www.ssrn.com/abstract=3398983.
93. Paul Krugman, "Opinion | Why Biden Will Need to Spend Big," *New York Times*, October 19, 2020, sec. Opinion, https://www.nytimes.com/2020/10/19/opinion/joe-biden-deficit-spending.html; see also Stiglitz, "The Economic Case for a Green New Deal," 100–105.
94. Chase Peterson-Withorn, "How Much Money America's Billionaires Have Made during the Covid-19 Pandemic," *Forbes*, April 30, 2021, https://www.forbes.com/sites/chasewithorn/2021/04/30/american-billionaires-have-gotten-12-trillion-richer-during-the-pandemic/.
95. United States Government Accountability Office, "Report to Congressional Addressees: Federal Reserve System; Opportunities Exist to Strengthen Policies and Processes for Managing Emergency Assistance" (Washington, D.C., 2011).
96. Christophe McGlade and Paul Ekins, "The Geographical Distribution of Fossil Fuels Unused When Limiting Global Warming to 2 °C,"

Nature 517, no. 7533 (January 2015): 187–90, https://doi.org/10.1038/nature14016; Dan Tong et al., "Committed Emissions from Existing Energy Infrastructure Jeopardize 1.5 °C Climate Target," *Nature 572*, no. 7769 (August 2019): 373–77, https://doi.org/10.1038/s41586-019-1364-3; Emily Grubert, "Fossil Electricity Retirement Deadlines for a Just Transition," *Science 370*, no. 6521 (December 4, 2020): 1171–73, https://doi.org/10.1126/science.abe0375; Dan Welsby et al., "Unextractable Fossil Fuels in a 1.5 °C World," *Nature 597*, no. 7875 (September 9, 2021): 230–34, https://doi.org/10.1038/s41586-021-03821-8; Kelly Trout et al., "Existing Fossil Fuel Extraction Would Warm the World beyond 1.5 °C," *Environmental Research Letters 17*, no. 6 (June 1, 2022): 064010, https://doi.org/10.1088/1748-9326/ac6228.

97. Gregor Semieniuk et al., "Stranded Fossil-Fuel Assets Translate to Major Losses for Investors in Advanced Economies," *Nature Climate Change 12*, no. 6 (June 2022): 532–38, https://doi.org/10.1038/s41558-022-01356-y; Jean Eaglesham and Vipal Monga, "Trillions in Assets May Be Left Stranded as Companies Address Climate Change," *Wall Street Journal*, November 20, 2021, sec. Markets, https://www.wsj.com/articles/trillions-in-assets-may-be-left-stranded-as-companies-address-climate-change-11637416980.
98. Gregor Semieniuk et al., "Potential Pension Fund Losses Should Not Deter High-Income Countries from Bold Climate Action," *Joule 7*, no. 7 (July 2023): 1383–87, https://doi.org/10.1016/j.joule.2023.05.023.
99. Ben Caldecott et al., "Stranded Assets: The Transition to a Low Carbon Economy, Overview for the Insurance Industry," *Lloyd's of London (Oxford Smith School, Oxford University,* 2017), 12; Alan Livsey, "Lex in Depth: The $900bn Cost of 'Stranded Energy Assets,'" February 4, 2020, https://www.ft.com/content/95efca74-4299-11ea-a43a-c4b328d9061c; see also James Leaton, "Unburnable Carbon: Are the World's Financial Markets Carrying a Carbon Bubble?" (London, UK: Carbon Tracker, July 13, 2011); J. Curtin et al., "Quantifying Stranding Risk for Fossil Fuel Assets and Implications for Renewable Energy Investment: A Review of the Literature," *Renewable and Sustainable Energy Reviews 116* (December 2019): 109402, https://doi.org/10.1016/j.rser.2019.109402.
100. See Jeff D. Colgan, Jessica F. Green, and Thomas N. Hale, "Asset Revaluation and the Existential Politics of Climate Change," *International Organization* 75, no. 2 (2021): 586–610, https://doi.org/10.1017/S0020818320000296.

101. International Energy Agency, "World Energy Investment 2023" (Paris, May 2023), 81, https://iea.blob.core.windows.net/assets/8834d3af-af60-4df0-9643-72e2684f7221/WorldEnergyInvestment2023.pdf.
102. UN Environment Programme, "2021 Report: The Production Gap," October 20, 2021, 14, https://productiongap.org/wp-content/uploads/2021/11/PGR2021_web_rev.pdf.
103. U.S. Bureau of Labor Statistics, "Industries at a Glance: Mining, Quarrying, and Oil and Gas Extraction: NAICS 21," accessed October 19, 2023, https://www.bls.gov/iag/tgs/iag21.htm; U.S. Bureau of Labor Statistics, "Industries at a Glance: Real Estate: NAICS 531," accessed October 19, 2023, https://www.bls.gov/iag/tgs/iag531.htm.
104. Sean O'Leary, "Appalachia's Natural Gas Counties: Contributing More to the U.S. Economy and Getting Less in Return" (*Ohio River Valley Institute,* February 2021).
105. Kate Aronoff, "Fossil Fuel Companies Are Job Killers," *The New Republic*, April 5, 2021, https://newrepublic.com/article/161937/fossil-fuel-companies-job-killers.
106. U.S. Bureau of Labor Statistics, "Solar Electric Power Generation – May 2020 OEWS Industry-Specific Occupational Employment and Wage Estimates," accessed April 28, 2021, https://www.bls.gov/oes/current/naics5_221114.htm.
107. U.S. Department of Energy, "United States Energy & Employment Report 2023," June 2023, 4, https://www.energy.gov/sites/default/files/2023-06/2023%20USEER%20REPORT-v2.pdf.
108. Noam Scheiber, "A Coal Miners Union Indicates It Will Accept a Switch to Renewable Energy in Exchange for Jobs.," *New York Times*, April 19, 2021, sec. Business, https://www.nytimes.com/live/2021/04/19/business/stock-market-today.

CHAPTER 3

1. The Heritage Foundation, "The Right Way to Ensure a Cleaner Environment," accessed October 28, 2023, https://www.heritage.org/environment/heritage-explains/the-right-way-ensure-cleaner-environment; Climate Investigations Center, "Heritage Foundation," accessed October 28, 2023, https://climateinvestigations.org/heritage-foundation/.
2. Bret Stephens, "Where My Climate Doubts Began to Melt," *New York Times*, October 28, 2022, sec. Opinion, https://www.nytimes.com/interactive/2022/10/28/opinion/climate-change-bret-stephens.html.

3. Ezra Klein, "Your Kids Are Not Doomed," *New York Times*, June 5, 2022, sec. Opinion, https://www.nytimes.com/2022/06/05/opinion/climate-change-should-you-have-kids.html.
4. Eric Levitz, "*Don't Look Up* Doesn't Get the Climate Crisis," *Intelligencer*, January 5, 2022, https://nymag.com/intelligencer/2022/01/dont-look-up-climate-metaphor-review.html, emphasis in the original.
5. Klein, "Your Kids Are Not Doomed."
6. Simon Kuznets, "Modern Economic Growth: Findings and Reflections," *NobelPrize.org,* 1971, https://www.nobelprize.org/prizes/economic-sciences/1971/kuznets/lecture/.
7. Robert Solow, "A Contribution to the Theory of Economic Growth," *The Quarterly Journal of Economics 70*, no. 1 (1956): 67.
8. Solow, 65.
9. Solow, 67.
10. Solow, 67.
11. Peter Passell, Marc Roberts, and Leonard Ross, "The Limits to Growth," *New York Times*, April 2, 1972, sec. Archives, https://www.nytimes.com/1972/04/02/archives/the-limits-to-growth-a-report-for-the-club-of-romes-project-on-the.html.
12. Carl Kaysen, "The Computer That Printed out W*O*L*F*," *Foreign Affairs 50*, no. 4 (1972): 660, https://doi.org/10.2307/20037939.
13. Robert Solow, "Is the End of the World at Hand?," in *The Economic Growth Controversy*, ed. Andrew Weintraub, Eli Schwartz, and J. Richard Aronson, Routledge Revivals (New York: Routledge, 2018), 49.
14. Solow, 51.
15. Solow, 48.
16. Robert Solow, "The Economics of Resources or the Resources of Economics (1974)," in *Economics of the Environment: Selected Readings*, ed. Robert Stavins, 7th Edition (Cheltenham, UK: Edward Elgar Publishing, 2019), 246.
17. Robert Solow, "Sustainability: An Economist's Perspective (1991)," in *Economics of the Environment: Selected Readings*, ed. Robert Stavins, 7th Edition (Cheltenham, UK: Edward Elgar Publishing, 2019), 426, 427.
18. Robert Solow, 427.
19. Solow, 427.
20. Paul Stegmann et al., "Plastic Futures and Their CO_2 Emissions," *Nature 612*, no. 7939 (December 8, 2022): 272–76, https://doi.org/10.1038/s41586-022-05422-5.
21. Solow, "Sustainability: An Economist's Perspective (1991)," 429.
22. Solow, 427; Solow, "Is the End of the World at Hand?," 246.

23. Quoted in Ross B. Emmett and Jesse Grabowski, "Better Lucky than Good: The Simon-Ehrlich Bet through the Lens of Financial Economics," *Ecological Economics 193* (March 2022): 2, https://doi.org/10.1016/j.ecolecon.2021.107322.
24. Emmett and Grabowski, 3, Table 1.
25. Solow, "Is the End of the World at Hand?," 51. According to the International Energy Agency, world oil production measured 1,940 million tonnes in 1971, 3,187 million in 1990, and 4,296 million in 2020. See International Energy Agency (IEA), "World Oil Production by Region, 1971–2020—Charts—Data & Statistics," IEA, accessed December 16, 2022, https://www.iea.org/data-and-statistics/charts/world-oil-production-by-region-1971-2020.
26. Branko Milanovic, "Western Money and Eastern Promises," *Global Inequality and More 3.0 (Substack newsletter)*, November 9, 2022, https://branko2f7.substack.com/p/western-money-and-eastern-promises.
27. See, for examples, International Monetary Fund, "Looming Ahead—Finance & Development, September 2014," accessed December 11, 2022, https://www.imf.org/external/pubs/ft/fandd/2014/09/nobels.htm; Washington Center for Equitable Growth, "In Conversation with Robert Solow," *Equitable Growth (blog)*, July 20, 2017, http://www.equitablegrowth.org/equitable-growth-in-conversation-robert-solow/.
28. Robert Solow, "Stray Thoughts on How It Might Go," in *In 100 Years: Leading Economists Predict the Future*, ed. Ignacio Palacios-Huerta (Cambridge, MA: MIT Press, 2015), 3.
29. George Akerlof et al., "Economists' Statement," *Climate Leadership Council*, January 16, 2019, https://clcouncil.org/economists-statement/.
30. Solow, "Is the End of the World at Hand?," 52.
31. Matto Mildenberger and Leah Stokes, "The Trouble with Carbon Pricing," *Boston Review*, September 23, 2020, http://bostonreview.net/science-nature-politics/matto-mildenberger-leah-c-stokes-trouble-carbon-pricing.
32. https://twitter.com/elonmusk/status/1533410745429413888
33. Mildenberger and Stokes, "The Trouble with Carbon Pricing."
34. Simon Black et al., "IMF Fossil Fuel Subsidies Data: 2023 Update," IMF, accessed October 29, 2023, https://www.imf.org/en/Publications/WP/Issues/2023/08/22/IMF-Fossil-Fuel-Subsidies-Data-2023-Update-537281.
35. Lida Weinstock, "How Climate Change May Affect the US Economy" (Washington, D.C.: Congressional Research Service, April 4, 2022).

36. Evan Herrnstadt and Terry Dinan, "CBO's Projection of the Effect of Climate Change on U.S. Economic Output," *Working Paper Series* (Washington, D.C.: Congressional Budget Office, September 2020).
37. Zeke Hausfather, "Analysis: When Might the World Exceed 1.5C and 2C of Global Warming?," *Carbon Brief*, December 4, 2020, https://www.carbonbrief.org/analysis-when-might-the-world-exceed-1-5c-and-2c-of-global-warming/.
38. Herrnstadt and Dinan, "CBO's Projection of the Effect of Climate Change on U.S. Economic Output," 8.
39. Matthew E. Kahn et al., "Long-Term Macroeconomic Effects of Climate Change: A Cross-Country Analysis," *Energy Economics 104* (December 2021): 7, https://doi.org/10.1016/j.eneco.2021.105624.
40. Holman Jenkins Jr., "Climate Change Is Affordable," *Wall Street Journal*, November 27, 2018, https://www.wsj.com/articles/climate-change-is-affordable-1543362461.
41. Michael Barbaro et al., "Broken Climate Pledges and Europe's Heat Wave," *New York Times*, July 19, 2022, sec. Podcasts, https://www.nytimes.com/2022/07/19/podcasts/the-daily/climate-change-europe-heat-wave-legislation.html, emphasis mine.
42. Joan Ballester et al., "Heat-Related Mortality in Europe during the Summer of 2022," *Nature Medicine 29*, no. 7 (July 2023): 1857–66, https://doi.org/10.1038/s41591-023-02419-z.
43. Riccardo Colacito, Bridget Hoffmann, and Toan Phan, "Temperature and Growth: A Panel Analysis of the United States," *Journal of Money, Credit and Banking 51*, no. 2–3 (March 2019): 314, https://doi.org/10.1111/jmcb.12574.
44. Colacito, Hoffmann, and Phan, 328.
45. William Nordhaus, "Reflections on the Economics of Climate Change," *Journal of Economic Perspectives 7*, no. 4 (November 1, 1993): 15, https://doi.org/10.1257/jep.7.4.11.
46. William Nordhaus, "Climate Change: The Ultimate Challenge for Economics," *American Economic Review 109*, no. 6 (June 1, 2019): 448, https://doi.org/10.1257/aer.109.6.1991.
47. Stephanie Condon, "Record-Breaking Heatwave Causes Cloud-Computing Problems," *ZDNET*, July 19, 2022, https://www.zdnet.com/article/the-uk-heat-wave-brings-down-some-oracle-and-google-cloud-data-centers/.
48. Eva Dou and Lyric Li, "China Shuts Factories, Rations Electricity as Heat Wave Stifles Economy," *Washington Post*, August 16, 2022, https://

www.washingtonpost.com/world/2022/08/16/china-heat-wave-climate-change-electricity/.

49. Marshall Burke, Solomon M. Hsiang, and Edward Miguel, "Global Non-Linear Effect of Temperature on Economic Production," *Nature* 527, no. 7577 (November 2015): 236, https://doi.org/10.1038/nature15725.
50. Burke, Hsiang, and Miguel, 235.
51. Burke, Hsiang, and Miguel, 237.
52. Marshall Burke and Vincent Tanutama, "Climatic Constraints on Aggregate Economic Output" (Cambridge, MA: National Bureau of Economic Research, April 2019), 5, https://doi.org/10.3386/w25779.
53. Burke, Hsiang, and Miguel, 237.
54. Zahra Hirji, "'Bad For Humans': The World Is on Track to Warm 3 Degrees Celsius This Century," *BuzzFeed News,* October 30, 2021, https://www.buzzfeednews.com/article/zahrahirji/global-warming-3-degrees-celsius-impact.
55. Timothy M. Lenton et al., "Quantifying the Human Cost of Global Warming," *Nature Sustainability* 6, no. 10 (May 22, 2023): 1239, https://doi.org/10.1038/s41893-023-01132-6.
56. Chi Xu et al., "Future of the Human Climate Niche," *Proceedings of the National Academy of Sciences* 117, no. 21 (May 26, 2020): 11350–55, https://doi.org/10.1073/pnas.1910114117.
57. Richard Newell, Brian Prest, and Steven Sexton, "The GDP-Temperature Relationship: Implications for Climate Change Damages," *Working Paper (Resources for the Future,* November 2018), 5.
58. Lida Weinstock, "How Climate Change May Affect the US Economy," 9, 10, emphasis mine.
59. Intergovernmental Panel on Climate Change, "Special Report on Climate Change and Land," 2019, Summary for Policymakers, 7.
60. UNICEF, "Nearly Half of All Child Deaths in Africa Stem from Hunger, Study Shows," *UNICEF Global Development Commons,* accessed December 17, 2022, https://gdc.unicef.org/resource/nearly-half-all-child-deaths-africa-stem-hunger-study-shows.
61. Intergovernmental Panel on Climate Change, "AR6 Climate Change 2021: Impacts, Adaptation, and Vulnerability," 2021, Technical Summary, 66.
62. Intergovernmental Panel on Climate Change, "Special Report on Climate Change and Land," 2019, Ch. 5, Executive Summary, 439.
63. Samuel Myers et al., "Current Guidance Underestimates Risk of Global Environmental Change to Food Security," *BMJ,* September 29, 2022, 1, 2, https://doi.org/10.1136/bmj-2022-071533.

64. Myers et al., 2.
65. Ariel Ortiz-Bobea et al., "Anthropogenic Climate Change Has Slowed Global Agricultural Productivity Growth," *Nature Climate Change 11*, no. 4 (April 2021): 306–12, https://doi.org/10.1038/s41558-021-01000-1.
66. World Meteorological Organization, "Meteorological and Humanitarian Agencies Sound Alert on East Africa," May 30, 2022, https://public.wmo.int/en/media/news/meteorological-and-humanitarian-agencies-sound-alert-east-africa.
67. United Nations, "Chad: Unprecedented Flooding Affects More than 340,000 People," *UN News*, August 26, 2022, https://news.un.org/en/story/2022/08/1125562; United Nations, "18 Million in Africa's Sahel on 'the Brink of Starvation,'" *UN News*, May 20, 2022, https://news.un.org/en/story/2022/05/1118702.
68. Ayesha Tandon, "West Africa's Deadly Rainfall in 2022 Made '80 Times More Likely' by Climate Change," *Carbon Brief*, November 16, 2022, https://www.carbonbrief.org/west-africas-deadly-rainfall-in-2022-made-80-times-more-likely-by-climate-change/.
69. The Editors, "Syria: Drought Puts Assad's 'Year of Wheat' in Peril," *Middle East Monitor (blog)*, June 21, 2021, https://www.middleeastmonitor.com/20210621-syria-drought-puts-assads-year-of-wheat-in-peril/.
70. Mustafa Sonmez, "Rise in Turkish Food Prices Sparks Fears of Shortages," *Al-Monitor*, June 11, 2021, https://www.al-monitor.com/originals/2021/06/rise-turkish-food-prices-sparks-fears-shortages.
71. Kapil Kajal, "In North India, Unforgiving Heatwaves Have Reduced the Yield and Quality of Wheat This Year," *Scroll.In*, June 9, 2022, https://scroll.in/article/1025744/the-heatwaves-roasting-north-india-have-reduced-the-yield-and-quality-of-its-wheat-this-year.
72. Amin Ahmed, "Festering Food Crisis Worries Red Cross; Guterres Terms Floods 'Unnatural,'" *DAWN.COM,* September 14, 2022, https://www.dawn.com/news/1710026.
73. Zhao Yimeng, "Agriculture Officials Work to Ensure Food Supply after Floods Damage Crops," *China Daily*, July 30, 2021, https://global.chinadaily.com.cn/a/202107/30/WS6103a913a310efa1bd6658fb.html.
74. Low Minmin, "'The Soil Is as Hard as Rock': Farmers Reel from China Heatwave as Food Inflation Looms," *Channel News Asia*, September 2, 2022, https://www.channelnewsasia.com/asia/china-heatwave-agriculture-food-inflation-chongqing-2915371.
75. Megan Durisin, "Smallest French Corn Crop since 1990 Shows Drought's Huge Toll," *Bloomberg.Com*, September 13, 2022, https://www.bloomberg.com/news/articles/2022-09-13/smallest-french-corn-crop-since-1990-shows-drought-s-huge-toll.

76. Aman Bhargava and Samuel Granados, "Europe's Driest Summer in 500 Years," *Reuters*, August 22, 2022, https://www.reuters.com/graphics/EUROPE-WEATHER/DROUGHT/jnvwenznyvw/.
77. Helena Horton, "Mass Crop Failures Expected in England as Farmers Demand Hosepipe Bans," *The Guardian*, August 12, 2022, sec. Environment, https://www.theguardian.com/environment/2022/aug/12/mass-crop-failures-expected-in-england-as-farmers-demand-hosepipe-bans; Madeleine Cuff, "UK Drought Shrinks Potatoes, Onions and Other Crops as Farmers Warn of Lasting Damage," *inews.co.uk,* August 12, 2022, https://inews.co.uk/news/uk-drought-farmers-struggle-feed-cattle-cheap-meat-heat wave-1793194.
78. Laura Reiley, "The Summer Drought's Hefty Toll on American Crops," *Washington Post*, September 7, 2022, https://www.washingtonpost.com/business/2022/09/05/crops-climate-drought-food/.
79. Kim Chipman, Zijia Song, and Tarso Veloso Ribeiro, "US Farmers Battle Floods, Heat in Bid to Replenish Food Supplies," *Bloomberg.Com*, June 29, 2022, https://www.bloomberg.com/news/articles/2022-06-29/how-extreme-weather-could-impact-crops-food-inflation; Chloe Sorvino, "Here's the Latest Data on Climate and Food and It's Not Good," *Forbes*, June 26, 2022, sec. Food & Drink, https://www.forbes.com/sites/chloesorvino/2022/06/26/heres-the-latest-data-on-climate-and-food-and-its-not-good/.
80. National Oceanic and Atmospheric Administration, "California," National Integrated Drought Information System, accessed December 17, 2022, https://www.drought.gov/states/california.
81. California Department of Water Resources, "Water Year 2023: Weather Whiplash, From Drought To Deluge," October 2023, 3, https://water.ca.gov/-/media/DWR%20Website/Web%20Pages/Water%20Basics/Drought/Files/Publications%20And%20Reports/Water%20Year%202023%20wrap%20up%20brochure_01?utm_medium=email&utm.
82. California Department of Water Resources, 10, 8.
83. California Department of Water Resources, "California's Groundwater Live," accessed December 17, 2022, https://sgma.water.ca.gov/CalGWLive/#stats; U.S. Department of the Interior, "Subsiding Areas in California | USGS California Water Science Center," accessed December 17, 2022, https://ca.water.usgs.gov/land_subsidence/california-subsidence-areas.html.

84. Dan Charles, "Without Enough Water to Go Around, Farmers in California Are Exhausting Aquifers," *NPR*, July 23, 2021, sec. Environment, https://www.npr.org/2021/07/22/1019483661/without-enough-water-to-go-around-farmers-in-california-are-exhausting-aquifers.
85. In 2021 the Central Valley Regional Water Quality Control Board released a report that reviewed the extant studies on the presence of the individual chemicals in the wastewater—including, among many others, known carcinogens like benzene and arsenic. The report found that 2021 levels of these chemicals would not harm human health, but an Inside Climate News investigation later revealed that Chevron, the biggest seller of wastewater to farms, is a regular client of GSI Environmental, the corporate "environmental research" firm who conducted the studies, and that their literature review did not examine the human-health effects of ingesting these chemicals together in a mixture, nor whether these chemicals would build up in the soil or the produce itself over time. See Katelyn Weisbrod, "California Regulators Banned Fracking Wastewater for Irrigation, but Allow Wastewater from Oil Drilling. Scientists Say There's Little Difference," *Inside Climate News (blog),* April 24, 2022, https://insideclimatenews.org/news/24042022/california-produced-water/.
86. Aruna Chandrasekhar and Anastasiia Zagoruichyk, "Commodity Profile: Wheat," *Carbon Brief*, August 8, 2022, https://interactive.carbonbrief.org/commodity-profile-wheat/.
87. Hannah Ritchie, "How Much of Global Greenhouse Gas Emissions Come from Food?," *Our World in Data*, accessed December 17, 2022, https://ourworldindata.org/greenhouse-gas-emissions-food.
88. Michael Grunwald, "No One Wants to Say 'Put Down That Burger,' but We Really Should," *New York Times*, December 15, 2022, https://www.nytimes.com/2022/12/15/opinion/food-diets-meat-biodiverstiy-cop15.html.
89. Jonathan Woetzel et al., "Will the World's Breadbaskets Become Less Reliable?," *McKinsey Global Institute*, May 2020.
90. Jonas Jägermeyr et al., "Climate Impacts on Global Agriculture Emerge Earlier in New Generation of Climate and Crop Models," *Nature Food 2*, no. 11 (November 1, 2021): 873–85, https://doi.org/10.1038/s43016-021-00400-y.
91. Solow, "Stray Thoughts on How It Might Go," 140.

CHAPTER 4

1. The Editorial Board, "John Kerry Lays It All Out on Climate Change," *Wall Street Journal*, January 19, 2023, https://www.wsj.com/articles/john-kerry-lays-it-all-out-on-climate-davos-11674170143.
2. The Editorial Board, "China's Coal Power Boom," *Wall Street Journal*, September 12, 2022, https://www.wsj.com/articles/chinas-coal-power-boom-beijing-xi-jinping-climate-energy-biden-administration-11650480857.
3. The Editorial Board, "China's Coal Power Boom."
4. Anthony Leiserowitz et al., "Politics and Global Warming, April 2022" (New Haven, CT: Yale Program on Climate Change Communication and George Mason University Center for Climate Change Communication, April 2022), https://climatecommunication.yale.edu/publications/politics-global-warming-april-2022/.
5. Michael Pollan, "Why Bother?," in *Global Environmental Politics: From Person to Planet*, ed. Simon Nicholson and Paul Kevin Wapner (Boulder: Paradigm Publishers, 2015), 289.
6. United Nations Development Program and Oxford Poverty and Human Development Initiative, "2022 Global Multidimensional Poverty Index," *Human Development Reports* (New York: United Nations, October 17, 2022), https://hdr.undp.org/content/2022-global-multidimensional-poverty-index-mpi.
7. UNFCCC, "Convention Documents," March 21, 1994, https://unfccc.int/process-and-meetings/the-convention/history-of-the-convention/convention-documents.
8. UNFCCC.
9. UNFCCC.
10. UNFCCC.
11. UNFCCC, "Paris Agreement (All Language Versions)," December 12, 2015, https://unfccc.int/process/conferences/pastconferences/paris-climate-change-conference-november-2015/paris-agreement.
12. Robert C. Byrd, "S.Res.98—105th Congress (1997–1998): A Resolution Expressing the Sense of the Senate Regarding the Conditions for the United States Becoming a Signatory to Any International Agreement on Greenhouse Gas Emissions under the United Nations Framework Convention on Climate Change," webpage, July 25, 1997, 1997/1998, https://www.congress.gov/bill/105th-congress/senate-resolution/98.
13. William Nordhaus, "To Slow or Not to Slow: The Economics of the Greenhouse Effect," *The Economic Journal 101*, no. 407 (1991): 920, https://doi.org/10.2307/2233864, emphasis mine.

14. Per Meilstrup, "The Runaway Summit: The Background Story of the Danish Presidency of COP15, the UN Climate Change Conference," in *Danish Foreign Policy Yearbook: 2010*, ed. Nanna Hvidt and Hans Mouritzen (Copenhagen: Danish Institute for International Studies, 2010), 125–28.
15. Krittivas Mukherjee and Gerard Wynn, "Big Developing States Reject Copenhagen Climate Plan," *Reuters*, December 2, 2009, sec. Environment, https://www.reuters.com/article/us-climate-idUSTRE5B14KC20091202.
16. Tobias Rapp, Christian Schwägerl, and Gerald Traufetter, "The Copenhagen Protocol: How China and India Sabotaged the UN Climate Summit," *Der Spiegel*, May 5, 2010, sec. International, https://www.spiegel.de/international/world/the-copenhagen-protocol-how-china-and-india-sabotaged-the-un-climate-summit-a-692861.html.
17. Rapp, Schwägerl, and Traufetter.
18. Meilstrup, "The Runaway Summit: The Background Story of the Danish Presidency of COP15, the UN Climate Change Conference," 132.
19. Meilstrup, 132.
20. Ian M. McGregor, "Disenfranchisement of Countries and Civil Society at COP15 in Copenhagen," *Global Environmental Politics 11*, no. 1 (February 2011): 4, https://doi.org/10.1162/GLEP_a_00039.
21. John Vidal, Allegra Stratton, and Suzanne Goldenberg, "Low Targets, Goals Dropped: Copenhagen Ends in Failure," *The Guardian*, December 19, 2009, sec. Environment, https://www.theguardian.com/environment/2009/dec/18/copenhagen-deal.
22. Meilstrup, "The Runaway Summit," 132; McGregor, "Disenfranchisement of Countries and Civil Society at COP15 in Copenhagen," 5.
23. McGregor, "Disenfranchisement of Countries and Civil Society at COP15 in Copenhagen," 5.
24. UNFCCC, "Copenhagen Accord," December 18, 2009, https://unfccc.int/resource/docs/2009/cop15/eng/l07.pdf.
25. See, e.g., Ed Henry, "Obama Announces Climate Change Deal with China, Other Nations," *CNN.com*, accessed June 12, 2023, http://www.cnn.com/2009/POLITICS/12/18/obama.copenhagen/index.html.
26. John Vidal, "Confidential Document Reveals Obama's Hardline US Climate Talk Strategy," *The Guardian*, April 12, 2010, sec. Environment, https://www.theguardian.com/environment/2010/apr/12/us-document-strategy-climate-talks.
27. Radoslav S. Dimitrov, "The Paris Agreement on Climate Change: Behind Closed Doors," *Global Environmental Politics 16*, no. 3 (August 2016): 3, https://doi.org/10.1162/GLEP_a_00361.

28. Dimitrov, 4.
29. John Vidal, "How a 'Typo' Nearly Derailed the Paris Climate Deal," *The Guardian*, December 16, 2015, sec. Environment, https://www.theguardian.com/environment/blog/2015/dec/16/how-a-typo-nearly-derailed-the-paris-climate-deal.
30. Daniel Bodansky, "The Paris Climate Change Agreement: A New Hope?," *American Journal of International Law 110*, no. 2 (April 2016): 294, https://doi.org/10.5305/amerjintelaw.110.2.0288; Dimitrov, 3.
31. Oil Change International, Earthworks, and The Center for International Environmental Law, "Permian Climate Bomb (Chapter 3)," *The Permian Climate Bomb*, 2021, https://www.permianclimatebomb.org/chapter-3; US Energy Information Administration, "Frequently Asked Questions: What Countries Are the Top Producers and Consumers of Oil?," 2022, https://www.eia.gov/tools/faqs/faq.php.
32. Valerie Richardson, "Obama Takes Credit for U.S. Oil-and-Gas Boom: 'That Was Me, People,'" *AP NEWS*, August 21, 2021, sec. Business, https://apnews.com/article/business-5dfbc1aa17701ae219239caad0bfefb2.
33. Oil Change International, "Investing in Disaster: Recent and Anticipated Final Investment Decisions for New Oil and Gas Production Beyond the 1.5°C Limit" (Washington, D.C., November 16, 2022), https://priceofoil.org/2022/11/16/investing-in-disaster/.
34. "Statement from President Joe Biden on Decision to Pause Pending Approvals of Liquefied Natural Gas Exports," The White House, accessed February 3, 2024, https://www.whitehouse.gov/briefing-room/statements-releases/2024/01/26/statement-from-president-joe-biden-on-decision-to-pause-pending-approvals-of-liquefied-natural-gas-exports/; Department of Fossil Energy and Carbon Management, "Summary of LNG Export Applications_12.11.23," accessed February 3, 2024, https://www.energy.gov/sites/default/files/2023-12/Summary%20of%20LNG%20Export%20Applications_12.11.23.pdf.
35. Curtis Williams, "US was Top LNG Exporter in 2023," *Reuters*, January 3, 2024, sec. Energy, https://www.reuters.com/business/energy/us-was-top-lng-exporter-2023-hit-record-levels-2024-01-02/.
36. Bill McKibben, "The Biden Administration's Next Big Climate Decision," *The New Yorker*, September 22, 2023, https://www.newyorker.com/news/daily-comment/the-biden-administrations-next-big-climate-decision.

37. International Energy Agency, "World Energy Investment 2023" (Paris, May 2023), https://iea.blob.core.windows.net/assets/8834d3af-af60-4df0-9643-72e2684f7221/WorldEnergyInvestment2023.pdf.
38. Rainforest Action Network and Banktrack, "Banking on Climate Chaos 2022" (California, March 29, 2022), https://www.bankingonclimatechaos.org/bankingonclimatechaos2022/.
39. Urgewald, "Press Release: Groundbreaking Research Reveals the Financiers of the Coal Industry," February 25, 2022, https://www.coalexit.org/sites/default/files/download_public/Financing%20GCEL%202020_Press%20Release_urgewald.pdf.
40. Jennifer Dlouhy, "G-7 Nations Tussle over Bid to Phase Out Coal Power by 2030," *Bloomberg*, April 10, 2023, https://www.bloomberg.com/news/articles/2023-04-10/g-7-nations-tussle-over-bid-to-phase-out-coal-power-by-2030.
41. Worldometer, "Coal Consumption by Country," accessed June 12, 2023, https://www.worldometers.info/coal/coal-consumption-by-country/.
42. Simon Evans et al., "COP26: Key Outcomes Agreed at the UN Climate Talks in Glasgow," *Carbon Brief*, November 15, 2021, https://www.carbonbrief.org/cop26-key-outcomes-agreed-at-the-un-climate-talks-in-glasgow/.
43. Evans et al.
44. Elizabeth Piper, William James, and Jake Spring, "India Criticises Fossil Fuel Language in COP26 Draft Deal," *Reuters*, November 13, 2021, sec. COP26, https://www.reuters.com/business/cop/india-says-consensus-over-cop26-climate-deal-remains-elusive-2021-11-13/.
45. Fiona Harvey and Rowena Mason, "Alok Sharma 'Deeply Frustrated' by India and China over Coal," *The Guardian*, November 14, 2021, sec. Environment, https://www.theguardian.com/environment/2021/nov/14/alok-sharma-deeply-frustrated-by-india-and-china-over-coal.
46. Jess Shankleman and Rathi Akshat, "India's Last-Minute Coal Defense at COP26 Hid Role of China, U.S.," *Bloomberg.Com*, November 13, 2021, https://www.bloomberg.com/news/articles/2021-11-13/india-s-last-minute-coal-defense-at-cop26-hid-role-of-china-u-s.
47. Jennifer Dlouhy, "US Backs Tough Fossil Fuel Phase Down Pledge at Climate Summit," *Bloomberg.Com*, November 16, 2022, https://www.bloomberg.com/news/articles/2022-11-16/us-backs-tough-fossil-fuel-phase-down-pledge-at-climate-summit.
48. Oxfam, "Climate Finance Short-Changed: The Real Value of the $100 Billion Commitment in 2019–2020" (Washington, D.C., October 20, 2022), https://www.oxfamamerica.org/explore/research-publications/

climate-finance-short-changed-the-real-value-of-the-100-billion-commitment-in-20192020/.

49. This $100 billion is separate from the Green Climate Fund that Obama also proposed in Copenhagen to provide grants and other concessional financing directly to projects in mitigation or adaptation. This fund seems to have been used mostly as a bribe to get Global South governments to sign the Copenhagen Accord. At the final plenary, the US delegate told the Parties in the auditorium that developed nations would not "operationalize" the Green Climate Fund unless the Accord was accepted. Like the Copenhagen Accord itself, this announcement was met with angry incredulity. Tuvalu, for example, strongly objected to what they called "being offered money to betray our people and our future," adding "our future is not for sale," to great applause throughout the hall, even though it was 3:15 a.m. at that point. But the US pressed ahead with this strategy to increase the number of the Accord's signatories. Among the diplomatic cables released by Wikileaks in 2010 were communications from the United States to other nations tying climate funding to their support, and the United States openly denied climate aid to both Bolivia and Ecuador after they declined to sign the Accord. In the end the Green Climate Fund was never fully capitalized anyway: to date the United States has provided $1 billion of Obama's initial $3 billion pledge. See McGregor, "Disenfranchisement of Countries and Civil Society at COP15 in Copenhagen"; Suzanne Goldenberg, "US Denies Climate Aid to Countries Opposing Copenhagen Accord," *The Guardian*, April 9, 2010, sec. Environment, https://www.theguardian.com/environment/2010/apr/09/us-climate-aid; Damian Carrington, "WikiLeaks Cables Reveal How US Manipulated Climate Accord," *The Guardian*, December 3, 2010, sec. Environment, https://www.theguardian.com/environment/2010/dec/03/wikileaks-us-manipulated-climate-accord; Jean Chemnick, "How Sending Climate Aid Abroad Helps the U.S.," *E&E News*, January 9, 2023, https://www.eenews.net/articles/how-sending-climate-aid-abroad-helps-the-u-s/.

50. Sara Schonhardt and Nick Sobczyk, "U.S. Climate Aid Pledge Runs into Capitol Hill Reality," *E&E News*, December 21, 2022, https://www.eenews.net/articles/u-s-climate-aid-pledge-runs-into-capitol-hill-reality/.

51. The White House, "Statement by President Trump on the Paris Climate Accord," June 1, 2017, https://trumpwhitehouse.archives.gov/briefings-statements/statement-president-trump-paris-climate-accord/.

52. Akshat Rathi and Archana Chaudhary, "India Wants $1 Trillion Before It Raises Targets to Cut Emissions," *Bloomberg.Com*, November 10, 2021, https://www.bloomberg.com/news/articles/2021-11-10/india-holds-back-on-climate-pledge-until-rich-nations-pay-1-trillion.
53. Xi Jinping, "Full Text of Xi Jinping's Speech at the China Communist Party Congress 2022," trans. Low De Wei, *Bloomberg.Com*, October 18, 2022, https://www.bloomberg.com/news/articles/2022-10-18/full-text-of-xi-jinping-s-speech-at-china-20th-party-congress-2022?sref=UIYAtP2Y.
54. Florence Tan and Chen Aizhu, "Saudi Aramco Boosts China Investment with Two Refinery Deals," *Reuters*, March 27, 2023, sec. Commodities News, https://www.reuters.com/article/saudi-aramco-china-refinery-idAFL1N35Z038.
55. Liu Lukun, "Gazprom Head Announces New LNG Flows to China," *China Daily*, March 29, 2023, https://www.chinadaily.com.cn/a/202303/29/WS642436b4a31057c47ebb7459.html.
56. The US-China Business Council, "China's Strategic Emerging Industries: Policy, Implementation, Challenges, and Recommendations" (Washington, D.C.: The US-China Business Council, March 2013), https://www.uschina.org/sites/default/files/sei-report.pdf.
57. Scott Moore, *China's Next Act: How Sustainability and Technology Are Reshaping China's Rise and the World's Future* (New York: Oxford University Press, 2022), 93.
58. Moore, 93.
59. International Energy Agency, "Energy Technology Perspectives 2023" (Paris: International Energy Agency, January 2023), https://iea.blob.core.windows.net/assets/a86b480e-2b03-4e25-bae1-da1395e0b620/EnergyTechnologyPerspectives2023.pdf.
60. Statista, "Global Polysilicon Manufacturing Capacity Share 2021," *Statista*, accessed June 13, 2023, https://www.statista.com/statistics/1334952/solar-polysilicon-manufacturing-capacity-share-by-country-or-region/; Dan Murtaugh, "China Mulls Protecting Solar Tech Dominance with Export Ban," *Bloomberg.Com*, January 26, 2023, https://www.bloomberg.com/news/articles/2023-01-26/china-mulls-protecting-solar-tech-dominance-with-export-ban.
61. Kate Aronoff, "To Fight Climate Change, Bring Back State Planning," *Intelligencer*, January 26, 2022, https://nymag.com/intelligencer/2022/01/to-fight-climate-change-bring-back-state-planning.html.
62. International Energy Agency, "Energy Technology Perspectives 2023."

63. Xiaoying You, "Analysis: What Does China's Coal Push Mean for Its Climate Goals?," *Carbon Brief*, March 29, 2022, https://www.carbonbrief.org/analysis-what-does-chinas-coal-push-mean-for-its-climate-goals/.
64. Lauri Myllyvirta, "The Increase in China's Power Generation...," *Tweet, Twitter*, February 13, 2023, https://twitter.com/laurimyllyvirta/status/1625118889909649411.
65. Bloomberg News, "China Added More Solar Panels in 2023 Than US Did In Its Entire History," *Bloomberg News*, January 6, 2024, sec. Green, https://www.bloomberg.com/news/articles/2024-01-26/china-added-more-solar-panels-in-2023-than-us-did-in-its-entire-history?sref=UIYAtP2Y.
66. The Communist Party of China, "Guiding Opinions on Coordinating and Strengthening the Work Related to Climate Change and Ecological Environmental Protection," *IEA*, accessed June 13, 2023, https://www.iea.org/policies/12989-guiding-opinions-on-coordinating-and-strengthening-the-work-related-to-climate-change-and-ecological-environmental-protection.
67. Hongqiao Liu, "5. The People's Bank of China (PBOC) Aims to "Comprehensively Factor Climate Change...," *Tweet, Twitter*, March 22, 2021, https://twitter.com/LHongqiao/status/1373925581625163778.
68. Ma Chenchen, "High Specification! This Central Leading Group Debuts Double-Carbon Work," *China Business News*, May 28, 2021, https://www.yicai.com/news/101064929.html.
69. Kelly Sims Gallagher and Xiaowei Xuan, *Titans of the Climate: Explaining Policy Process in the United States and China, American and Comparative Environmental Policy* (Cambridge, MA: The MIT Press, 2018), 98.
70. The Center for Research on Energy and Clean Air, "BRIEFING: 12.8 GW of Chinese Overseas Coal Projects Cancelled, but 19 GW Could Still Go Ahead," April 2022, https://energyandcleanair.org/wp/wp-content/uploads/2022/04/Final_Chinese-overseas-briefing_April2022.pdf.
71. The State Council of China, "Full Text: Action Plan for Carbon Dioxide Peaking Before 2030," accessed June 13, 2023, http://www.news.cn/english/2021-10/27/c_1310270985.htm.
72. The Communist Party of China Central Committee, "Full Text: Working Guidance for Carbon Dioxide Peaking and Carbon Neutrality in Full and Faithful Implementation of the New Development Philosophy," accessed June 13, 2023, http://www.news.cn/english/2021-10/24/c_

1310265726.htm. All further quotations of this text will be taken from this reference.

73. Eurostat, "Renewable Energy Statistics," accessed June 13, 2023, https://ec.europa.eu/eurostat/statistics-explained/index.php?title=Renewable_energy_statistics; Office of Energy Efficiency and Renewable Energy, "Renewable Energy," Energy.gov, accessed June 13, 2023, https://www.energy.gov/eere/renewable-energy.
74. Global Energy Monitor et al., "Boom and Bust Coal: Tracking the Global Coal Plant Pipeline" (*Global Energy Monitor,* April 25, 2022), https://globalenergymonitor.org/report/boom-and-bust-coal-2022/.
75. Centre for Research on Energy and Global Energy Monitor, "China Permits Two New Coal Power Plants Per Week in 2022" (*Centre for Research on Energy and Clean Air,* February 27, 2023), https://energyandcleanair.org/publication/china-permits-two-new-coal-power-plants-per-week-in-2022/.
76. Amy Hawkins, "China Ramps Up Coal Power Despite Carbon Neutral Pledges," *The Guardian*, April 24, 2023, sec. World news, https://www.theguardian.com/world/2023/apr/24/china-ramps-up-coal-power-despite-carbon-neutral-pledges.
77. International Energy Agency, "Production—Coal Information: Overview—Analysis," *IEA*, accessed June 14, 2023, https://www.iea.org/reports/coal-information-overview/production.
78. Xiaoying You, "Analysis: How Power Shortages Might 'Accelerate' China's Climate Action," *Carbon Brief*, October 13, 2021, https://www.carbonbrief.org/analysis-how-power-shortages-might-accelerate-chinas-climate-action/.
79. David Sandlow et al., "Guide to Chinese Climate Policy 2022" (Oxford, UK: The Oxford Institute for Energy Studies, October 14, 2022), 49.
80. The State Council of China, "Full Text."
81. Nicolas Véron, "Much of the Global South Is on Ukraine's Side," *Peterson Institute for International Economics,* March 8, 2023, https://www.piie.com/blogs/realtime-economics/much-global-south-ukraines-side.
82. The Center for Research on Energy and Clean Air, "China's Climate Transition Outlook, 2022," November 2022, https://energyandcleanair.org/wp/wp-content/uploads/2022/11/Chinas-Climate-Transition_Outlook-2022.pdf.
83. International Energy Agency, "Datafile: World Energy Investment 2022," *IEA*, accessed June 14, 2023, https://www.iea.org/data-and-statistics/data-product/world-energy-investment-2022-datafile.

84. Andrew Hayley, "China's Installed Non-Fossil Fuel Electricity Capacity Exceeds 50% of Total," *Reuters*, June 12, 2023, https://www.reuters.com/business/energy/chinas-installed-non-fossil-fuel-electricity-capacity-exceeds-50-total-2023-06-12/; U.S. Energy Information Administration (EIA), "FAQ: What Is U.S. Electricity Generation by Energy Source?," accessed June 14, 2023, https://www.eia.gov/tools/faqs/faq.php.
85. For the "fog of enactment," see Leah Cardamore Stokes, *Short Circuiting Policy: Interest Groups and the Battle over Clean Energy and Climate Policy in the American States, Studies in Postwar American Political Development* (New York: Oxford University Press, 2020).
86. See Chapters 1 and 3 in Kenneth Lieberthal and Michel Oksenberg, *Policy Making in China: Leaders, Structures, and Processes* (Princeton, NJ: Princeton University Press, 1988).
87. Sam Geall and Adrian Ely, "Narratives and Pathways towards an Ecological Civilization in Contemporary China," *The China Quarterly 236* (December 2018): 1184, https://doi.org/10.1017/S0305741018001315.
88. Gallagher and Xuan, *Titans of the Climate*, 53; David Sandlow et al., "Guide to Chinese Climate Policy 2022," 8, 44.
89. Yuen Yuen Ang, "How Beijing Commands: Grey, Black, and Red Directives from Deng to Xi," *The China Quarterly*, forthcoming.
90. See Alex Wang, "Symbolic Legitimacy and Chinese Environmental Reform," *International Law 48*, no. 4 (2018): 699–760.
91. Zheng Zeguang, "China Will Honour Its Climate Pledges – Look at the Changes We Have Already Made," *The Guardian*, October 27, 2021, sec. Opinion, https://www.theguardian.com/commentisfree/2021/oct/27/china-climate-pledges-cop26-emissions.
92. Wang, "Symbolic Legitimacy and Chinese Environmental Reform."
93. Kripa Jayaram, Chris Kay, and Dan Murtaugh, "China's Clean Air Campaign Is Bringing Down Global Pollution," *Bloomberg*, June 14, 2022, https://www.bloomberg.com/news/articles/2022-06-14/china-s-clean-air-campaign-is-bringing-down-global-pollution?sref=UIYAtP2Y.
94. Centre for Research on Energy and Clean Air, "China's Climate Transition: Outlook 2022" (*Centre for Research on Energy and Clean Air,* November 2022).
95. Yale Program on Climate Change Communication, "Climate Change in the Chinese Mind," https://www.youtube.com/watch?v=LTM_CPja81w.

96. Jinping, "Full Text of Xi Jinping's Speech at the China Communist Party Congress 2022."
97. Rebecca Beitsch, "National Security Adviser: US Needs to Get 'Own House in Order' to Strengthen Position Abroad," *The Hill*, January 29, 2021, https://thehill.com/policy/national-security/536499-sullivan-stresses-work-at-home-to-strengthen-us-position-abroad/.
98. Zach Coleman and Karl Mathiesen, "Newest Cause for Climate Optimism? The U.S. Rivalry with China," *POLITICO*, August 20, 2022, https://www.politico.com/news/2022/08/20/china-clean-energy-ira-climate-00052684.
99. Somini Sengupta, "The Shift to Renewable Energy Is Speeding Up. Here's How," *New York Times*, January 31, 2023, sec. Climate, https://www.nytimes.com/2023/01/31/climate/renewable-energy-transition.html.
100. See Fergus Green, "The Logic of Fossil Fuel Bans," *Nature Climate Change 8*, no. 6 (June 2018): 449–51, https://doi.org/10.1038/s41558-018-0172-3; Greg Muttitt and Sivan Kartha, "Equity, Climate Justice and Fossil Fuel Extraction: Principles for a Managed Phase Out," *Climate Policy 20*, no. 8 (September 13, 2020): 1024–42, https://doi.org/10.1080/14693062.2020.1763900; Boyan Yanovski and Kai Lessmann, "Financing the Fossil Fuel Phase-Out," *SSRN Electronic Journal*, 2021, https://doi.org/10.2139/ssrn.3903026; Mark Paul and Lina Moe, "An Economist's Case for Restrictive Supply Side Policies: Ten Policies to Manage the Fossil Fuel Transition" (Washington, D.C.: Climate and Community Project, March 2023); "Resources, Research & Publications," The Fossil Fuel Non-Proliferation Treaty Initiative, accessed October 31, 2023, https://fossilfueltreaty.org/resources.
101. Emma Dumain, "74% of Registered Voters Favor Carbon Tariff," *E&E Daily*, June 21, 2023, https://subscriber.politicopro.com/article/eenews/2023/06/21/poll-74-of-registered-voters-favor-carbon-tariff-00102783.
102. Jason Bordoff and Meghan L. O'Sullivan, "Green Upheaval," *Foreign Affairs*, November 30, 2021, https://www.foreignaffairs.com/articles/world/2021-11-30/geopolitics-energy-green-upheaval.
103. Senator Mitt Romney [@SenatorRomney], "The Critical Minerals Required for Batteries & Solar Panels Largely Come from China Because We're Not Mining Enough Here at Home. We Cannot Afford for the CCP to Be the Global Leader in Energy Production—We Need a Strategy for How We Will Lead the World in the New Energy Economy. Https://T.Co/bqkVFnLuL3," Tweet, Twitter, December

9, 2022, https://twitter.com/SenatorRomney/status/1601279117252386816.

104. Corbin Hiar, "Some Republicans See Climate Danger. They Voted 'No' Anyway," *E&E News*, August 12, 2022, https://subscriber.politicopro.com/article/eenews/2022/08/12/some-republicans-see-climate-dangers-they-voted-no-anyway-00050830.

CHAPTER 5

1. Coral Davenport and Lisa Friedman, "The Battle Lines Are Forming in Biden's Climate Push," *New York Times*, January 26, 2021, sec. Climate, https://www.nytimes.com/2021/01/26/climate/biden-climate-change.html.
2. "President Trump Signs Barrasso's Bipartisan Carbon Capture Bill into Law," *U.S. Senate Committee on Environment and Public Works,* accessed March 9, 2022, https://www.epw.senate.gov/public/index.cfm/2020/12/president-trump-signs-barrasso-s-bipartisan-carbon-capture-bill-into-law.
3. "Carbon Dioxide Removal Mission – Mission Innovation," accessed March 9, 2022, http://mission-innovation.net/missions/carbon-dioxide-removal/.
4. Quoted in Rachel Koning Beals, "Biden's Climate Envoy Kerry Gives Natural Gas a 10-Year Expiration Date," *MarketWatch*, accessed November 2, 2023, https://www.marketwatch.com/story/bidens-climate-envoy-kerry-gives-natural-gas-a-10-year-expiration-date-11650567971.
5. International Energy Agency, "Bioenergy with Carbon Capture and Storage," accessed November 1, 2023, https://www.iea.org/energy-system/carbon-capture-utilisation-and-storage/bioenergy-with-carbon-capture-and-storage.
6. Karin Rives, "Only Still-Operating Carbon Capture Project Battled Technical Issues in 2021," *S & P Global Market Intelligence*, January 6, 2022, https://www.spglobal.com/marketintelligence/en/news-insights/latest-news-headlines/only-still-operating-carbon-capture-project-battled-technical-issues-in-2021-68302671.
7. Geoff Leo, "Carbon Capture Facility by SNC-Lavalin Has 'Serious Design Issues,'" *CBC News*, October 27, 2015, https://www.cbc.ca/news/canada/saskatchewan/snc-lavalin-carbon-capture-project-saskpower-1.3291554; Stefani Langenegger, "Carbon Capture Coal Plants Double the Price of Power," *CBC News*, June 17, 2016, https://www.

cbc.ca/news/canada/saskatchewan/carbon-capture-power-prices-1.3641066.

8. Andy Rowell, "Carbon Capture: Five Decades of False Hope, Hype, and Hot Air" (Washington, D.C.: Oil Change International, June 17, 2021), https://priceofoil.org/2021/06/17/carbon-capture-five-decades-of-industry-false-hope-hype-and-hot-air/.
9. Chris Mooney, "America's First 'Clean Coal' Plant Is Now Operational—and Another Is on the Way," *Washington Post*, January 10, 2017, https://www.washingtonpost.com/news/energy-environment/wp/2017/01/10/americas-first-clean-coal-plant-is-now-operational-and-another-is-on-the-way/; John Schwartz, "Can Carbon Capture Technology Prosper under Trump?," *New York Times*, January 2, 2017, sec. Science, https://www.nytimes.com/2017/01/02/science/donald-trump-carbon-capture-clean-coal.html.
10. Nichola Groom, "Problems Plagued U.S. CO2 Capture Project before Shutdown," *Reuters*, August 6, 2020, sec. Environment, https://www.reuters.com/article/us-usa-energy-carbon-capture-idUSKCN2523K8.
11. June Sekera and Andreas Lichtenberger, "Assessing Carbon Capture: Public Policy, Science, and Societal Need: A Review of the Literature on Industrial Carbon Removal," *Biophysical Economics and Sustainability* 5, no. 3 (September 2020): 13, https://doi.org/10.1007/s41247-020-00080-5.
12. MIT, "Carbon Capture and Sequestration Technologies," *MIT Carbon Capture and Sequestration Technologies*, accessed March 9, 2022, https://sequestration.mit.edu/tools/projects/quest.html; Anya Zoledziowski, "Shell's Massive Carbon Capture Plant Is Emitting More Than It's Capturing," *Vice News*, January 20, 2022.
13. Anmar Frangoul, "'For Us, It Is Not a Solution': Enel CEO Skeptical over the Use of Carbon Capture," *CNBC*, November 25, 2021, https://www.cnbc.com/2021/11/25/climate-enel-ceo-skeptical-of-carbon-capture-and-storage-technology.html.
14. National Academies of Sciences, Engineering, and Medicine, *Negative Emissions Technologies and Reliable Sequestration: A Research Agenda* (Washington, D.C.: National Academies Press, 2019), 165–66, https://doi.org/10.17226/25259.
15. Jay Fuhrman et al., "Food–Energy–Water Implications of Negative Emissions Technologies in a +1.5 °C Future," *Nature Climate Change 10*, no. 10 (October 2020): 12, https://doi.org/10.1038/s41558-020-0876-z.
16. Clair Gough et al., "Challenges to the Use of BECCS as a Keystone Technology in Pursuit of 1.50C," *Global Sustainability 1* (2018): 3, https://doi.org/10.1017/sus.2018.3.

17. Naomi E. Vaughan and Clair Gough, "Expert Assessment Concludes Negative Emissions Scenarios May Not Deliver," *Environmental Research Letters 11*, no. 9 (September 1, 2016): 095003, https://doi.org/10.1088/1748-9326/11/9/095003.
18. International Energy Agency, "Direct Air Capture," *IEA*, accessed November 1, 2023, https://www.iea.org/energy-system/carbon-capture-utilisation-and-storage/direct-air-capture.
19. International Energy Agency.
20. National Academies of Sciences, Engineering, and Medicine, *Negative Emissions Technologies and Reliable Sequestration*, 228.
21. National Academies of Sciences, Engineering, and Medicine, 190.
22. Sabine Fuss et al., "Negative Emissions—Part 2: Costs, Potentials and Side Effects," *Environmental Research Letters 13*, no. 6 (June 1, 2018): 18, https://doi.org/10.1088/1748-9326/aabf9f.
23. Sudipta Chatterjee and Kuo-Wei Huang, "Unrealistic Energy and Materials Requirement for Direct Air Capture in Deep Mitigation Pathways," *Nature Communications 11*, no. 1 (December 2020): 3287, https://doi.org/10.1038/s41467-020-17203-7.
24. Ángel Galán-Martín et al., "Delaying Carbon Dioxide Removal in the European Union Puts Climate Targets at Risk," *Nature Communications 12*, no. 1 (December 2021): 6490, https://doi.org/10.1038/s41467-021-26680-3.
25. Jan C. Minx et al., "Negative Emissions—Part 1: Research Landscape and Synthesis," *Environmental Research Letters 13*, no. 6 (June 1, 2018): 13, https://doi.org/10.1088/1748-9326/aabf9b; Fuss et al., "Negative Emissions—Part 2," 13.
26. National Academies of Sciences, Engineering, and Medicine, *Negative Emissions Technologies and Reliable Sequestration*, 353, 360.
27. European Academies Science Advisory Council, "Negative Emission Technologies: What Role in Meeting Paris Agreement Targets?," February 2018, 1, https://easac.eu/fileadmin/PDF_s/reports_statements/Negative_Carbon/EASAC_Report_on_Negative_Emission_Technologies.pdf.
28. Intergovernmental Panel on Climate Change, "Climate Change 2021: The Physical Science Basis," 2021, 775, https://report.ipcc.ch/ar6/wg1/IPCC_AR6_WGI_FullReport.pdf.
29. Intergovernmental Panel on Climate Change, "Global Warming of 1.5°C," 2018, Ch. 2, 96, https://www.ipcc.ch/sr15/.

30. ExxonMobil, "CCS in Action," ExxonMobil, accessed March 10, 2022, https://corporate.exxonmobil.com:443/Climate-solutions/CCS-in-action.
31. BP, "CCS and Hydrogen | Who We Are | Home," bp United States, accessed November 1, 2023, https://www.bp.com/en_us/united-states/home/who-we-are/advocating-for-net-zero-in-the-us/ccs-and-hydrogen.html.
32. Chevron, "Gorgon Carbon Capture and Storage," chevron.com, accessed March 10, 2022, https://australia.chevron.com/our-businesses/gorgon-project/carbon-capture-and-storage.
33. See, for example, ExxonMobil, "ExxonMobil Announces Ambition for Net Zero Greenhouse Gas Emissions by 2050," ExxonMobil, accessed November 1, 2023, https://corporate.exxonmobil.com/news/news-releases/2022/0118_exxonmobil-announces-ambition-for-net-zero-greenhouse-gas-emissions-by-2050.
34. BP, "BP Sets Ambition for Net Zero by 2050, Fundamentally Changing Organisation to Deliver," bp global, accessed March 10, 2022, https://www.bp.com/en/global/corporate/news-and-insights/press-releases/bernard-looney-announces-new-ambition-for-bp.html.
35. Shell, "Our Climate Target," accessed March 10, 2022, https://www.shell.com/energy-and-innovation/the-energy-future/our-climate-target.html.
36. Oxy, "Occidental Petroleum Corporation – Climate Change 2022," n.d., https://www.oxy.com/siteassets/documents/sustainability/Oxy_CDP_Climate_Change.pdf.
37. Oxy, "Occidental Enters into Agreement to Acquire Direct Air Capture Technology Innovator Carbon Engineering," accessed November 1, 2023, https://www.oxy.com/news/news-releases/occidental-enters-into-agreement-to-acquire-direct-air-capture-technology-innovator-carbon-engineering/.
38. Molly Taft, "Big Oil's New Slate of Buzzwords Are Greenwashing at Its Worst," *Gizmodo*, March 3, 2021, https://gizmodo.com/big-oil-s-new-slate-of-desperate-buzzwords-are-greenwas-1846395653?utm_content=earther&utm_source=twitter&utm_medium=SocialMarketing&utm_campaign=dlvrit; Amy Harder, "Oil Giant Occidental Is Betting on Becoming a 'Carbon Management Company,'" Axios, December 4, 2020, https://www.axios.com/occidental-petroleum-oil-companies-energy-climate-aba58815-c941-4843-9cbd-821f916947b6.html.

39. ExxonMobil, "Advancing Climate Solutions 2022 Progress Report," 2022, 7, https://static1.squarespace.com/static/61b952aedd1042553 3c44259/t/63cfc7cabc68a5418b4d2d0e/1674561501986/exxonmobil-advancing-climate-solutions-2022-progress-report+%281%29.pdf.
40. Giulia Realmonte et al., "An Inter-Model Assessment of the Role of Direct Air Capture in Deep Mitigation Pathways," *Nature Communications 10*, no. 1 (December 2019), 2, https://doi.org/10.1038/s41467-019-10842-5.
41. Minx et al., "Negative Emissions—Part 1."
42. International Energy Agency, "World Energy Investment 2023" (Paris, May 2023), 81, 67 https://iea.blob.core.windows.net/assets/8834d3af-af60-4df0-9643-72e2684f7221/WorldEnergyInvestment2023.pdf.
43. InfluenceMap, "Big Oil's Real Agenda on Climate Change" https://influencemap.org/report/How-Big-Oil-Continues-to-Oppose-the-Paris-Agreement-38212275958aa21196dae3b76220bddc; see also Robert J. Brulle, "The Climate Lobby: A Sectoral Analysis of Lobbying Spending on Climate Change in the USA, 2000 to 2016," *Climatic Change 149*, no. 3–4 (August 2018): 289–303, https://doi.org/10.1007/s10584-018-2241-z.
44. Hiroko Tabuchi, "In Your Facebook Feed: Oil Industry Pushback Against Biden Climate Plans" *New York Times*, September 30, 2021, https://www.nytimes.com/2021/09/30/climate/api-exxon-biden-climate-bill.html.
45. Olivia Azadegan, "Common Concerns Raised about Carbon Capture and Storage Technology" (Boston, MA: The Clean Air Task Force, September 22, 2021), 3, https://cdn.catf.us/wp-content/uploads/2021/10/21091658/CAFF_CCSMythBusting_Proof_09.22.21_v2-3.pdf.
46. S. Julio Friedmann, "Engineered CO_2 Removal, Climate Restoration, and Humility," *Frontiers in Climate 1* (July 26, 2019): 1, 4, https://doi.org/10.3389/fclim.2019.00003.
47. Climeworks, "Carbon Capture: A Critical Tool in the Climate Restoration Toolbox," *Grist*, June 1, 2021, https://grist.org/article/carbon-capture-a-critical-tool-in-the-climate-restoration-toolbox/.
48. Intergovernmental Panel on Climate Change, "Climate Change 2021: The Physical Science Basis," Summary for Policymakers, 20.
49. Intergovernmental Panel on Climate Change, "Climate Change 2022: Impacts, Adaptation, and Vulnerability," 2022, Summary for Policymakers, 20, https://report.ipcc.ch/ar6/wg2/IPCC_AR6_WGII_FullReport.pdf.

50. Xinru Li et al., "Irreversibility of Marine Climate Change Impacts under Carbon Dioxide Removal," *Geophysical Research Letters 47*, no. 17 (September 16, 2020), https://doi.org/10.1029/2020GL088507; Katarzyna B. Tokarska and Kirsten Zickfeld, "The Effectiveness of Net Negative Carbon Dioxide Emissions in Reversing Anthropogenic Climate Change," *Environmental Research Letters 10*, no. 9 (September 1, 2015): 094013, https://doi.org/10.1088/1748-9326/10/9/094013.
51. Kirsten Zickfeld et al., "Asymmetry in the Climate–Carbon Cycle Response to Positive and Negative CO_2 Emissions," *Nature Climate Change 11*, no. 7 (July 2021): 613–17, https://doi.org/10.1038/s41558-021-01061-2.
52. Holly Buck, "We Need to Change How We Talk about Climate Action," *Jacobin*, May 22, 2021, https://jacobinmag.com/2021/05/climate-change-green-new-deal-technology.
53. John Schwartz, "Can Carbon Capture Technology Prosper under Trump?," *New York Times,* January 2, 2017, sec. Science, https://www.nytimes.com/2017/01/02/science/donald-trump-carbon-capture-clean-coal.html.
54. Buck, "We Need to Change How We Talk about Climate Action."
55. Cathy Bussewitz, "Insider Q&A: Occidental Wants to Be the Tesla of Carbon Capture," *ABC News*, March 21, 2021, https://abcnews.go.com/Business/wireStory/insider-qa-occidental-tesla-carbon-capture-76589343.
56. Buck, "We Need to Change How We Talk about Climate Action."
57. Buck.
58. Holly Jean Buck, *Ending Fossil Fuels: Why Net Zero Is Not Enough* (Brooklyn: Verso Books, 2021), 145.
59. Buck, 177.
60. Buck, "We Need to Change How We Talk about Climate Action."
61. Gates wrote in his 2021 climate book that "the DAC-based approach is really just a thought experiment" which "almost certainly can't be developed and deployed quickly enough to prevent dire harm to the environment"—and that in any case "it's not clear that we could store hundreds of billions of tons of carbon safely" (which is correct)—but he seemed to subsequently change his mind. See Bill Gates, *How to Avoid a Climate Disaster: The Solutions We Have and the Breakthroughs We Need*, First edition (New York: Alfred A. Knopf, 2021), 64.
62. Dale Jamieson, "Ethics and Intentional Climate Change," *Climatic Change 33*, no. 3 (July 1996): 333, https://doi.org/10.1007/BF00142580.

63. Phil Renforth and Jennifer Wilcox, "Editorial: The Role of Negative Emission Technologies in Addressing Our Climate Goals," *Frontiers in Climate 2* (January 28, 2020): 1, https://doi.org/10.3389/fclim.2020.00001.
64. Victoria Campbell-Arvai et al., "The Influence of Learning about Carbon Dioxide Removal (CDR) on Support for Mitigation Policies," *Climatic Change 143*, no. 3–4 (August 2017): 321–36, https://doi.org/10.1007/s10584-017-2005-1.
65. Corbin Hiar, "Elon Musk Joins Race to Bury CO2: 'The Math Backs That Up,'" *E&E News*, January 5, 2021, https://subscriber.politicopro.com/article/eenews/1063723359.
66. US Department of Energy, "Interagency Task Force on Carbon Capture and Storage," *Energy.gov,* accessed March 11, 2022, https://www.energy.gov/fecm/services/advisory-committees/interagency-task-force-carbon-capture-and-storage.
67. Ahmed Abdulla et al., "Explaining Successful and Failed Investments in U.S. Carbon Capture and Storage Using Empirical and Expert Assessments," *Environmental Research Letters 16*, no. 1 (January 1, 2021): 014036, https://doi.org/10.1088/1748-9326/abd19e.
68. Guloren Turan, Alex Zapantis, and David Kearns, "Global Status of CCS: 2021" (Melbourne, Australia: Global CCS Institute, 2021).
69. Joe Manchin, "Manchin Statement on Build Back Better Act," *The Official U.S. Senate website of Senator Joe Manchin of West Virginia,* December 19, 2021, https://www.manchin.senate.gov/newsroom/press-releases/manchin-statement-on-build-back-better-act.
70. "Testimony of Mr. Brad Crabtree Director, Carbon Capture Coalition Before the House Select Committee on the Climate Crisis" (Washington, D.C., September 26, 2019).
71. Food and Water Watch, "Biden Climate Watch," *Food & Water Watch,* June 30, 2021, https://www.foodandwaterwatch.org/2021/06/30/biden-cabinet-climate/.
72. Food and Water Watch.
73. Oxy, "1PointFive Selected for U.S. Department of Energy Grant to Develop South Texas Direct Air Capture Hub," accessed November 2, 2023, https://www.oxy.com/news/news-releases/1pointfive-selected-for-u.s.-department-of-energy-grant-to-develop-south-texas-direct-air-capture-hub/.
74. "Carbon Dioxide Removal," Center for Climate and Energy Solutions, accessed November 2, 2023, https://www.c2es.org/content/carbon-dioxide-removal/.

75. Intergovernmental Panel on Climate Change, "Statement on IPCC Principles and Procedures—IPCC," IPCC, accessed March 11, 2022, https://www.ipcc.ch/2010/02/02/statement-on-ipcc-principles-and-procedures/.
76. Sean Low and Stefan Schäfer, "Is Bio-Energy Carbon Capture and Storage (BECCS) Feasible? The Contested Authority of Integrated Assessment Modeling," *Energy Research & Social Science 60* (February 2020): 101326, https://doi.org/10.1016/j.erss.2019.101326.
77. Benjamin Storrow, "IPCC Modelers' Secret Weapon: Negative Emissions Tech," *E&E News*, December 22, 2020, https://subscriber.politicopro.com/article/eenews/1063721299. It would behoove these modelers, however, to do more outreach reiterating that their work is not prescriptive. Because, of course, policymakers view modeled mitigation trajectories as maps guiding them into the future. A 2021 Congressional Research Service report on climate scenario modeling, for instance, advises that models may "provide a foundation for Members of Congress who are considering climate change mitigation proposals" since they are "specifically designed to find technology deployments that meet specified climate or emissions constraints, typically in a lowest-cost manner." See Michael Westphal, "Greenhouse Gas Emissions Scenarios: Background, Issues, and Policy Relevance" (Washington, D.C.: Congressional Research Service, June 3, 2021).
78. Intergovernmental Panel on Climate Change, "Climate Change 2022: Impacts, Adaptation, and Vulnerability," Ch. 2, 100.
79. Neil Grant et al., "Cost Reductions in Renewables Can Substantially Erode the Value of Carbon Capture and Storage in Mitigation Pathways," *One Earth 4*, no. 11 (November 2021): 1588–1601, https://doi.org/10.1016/j.oneear.2021.10.024.
80. Realmonte et al., "An Inter-Model Assessment of the Role of Direct Air Capture in Deep Mitigation Pathways," 9.
81. Grant et al., "Cost Reductions in Renewables Can Substantially Erode the Value of Carbon Capture and Storage in Mitigation Pathways."
82. Joeri Rogelj et al., "A New Scenario Logic for the Paris Agreement Long-Term Temperature Goal," *Nature 573*, no. 7774 (September 19, 2019): 357, https://doi.org/10.1038/s41586-019-1541-4; see also Joeri Rogelj et al., "Scenarios towards Limiting Global Mean Temperature Increase below 1.5 °C," *Nature Climate Change 8*, no. 4 (April 2018): 325–32, https://doi.org/10.1038/s41558-018-0091-3; Keywan Riahi et al., "The Shared Socioeconomic Pathways and Their Energy, Land Use, and Greenhouse Gas Emissions Implications: An Overview,"

Global Environmental Change 42 (January 2017): 153–68, https://doi.org/10.1016/j.gloenvcha.2016.05.009.

83. Kevin Anderson and Glen Peters, "The Trouble with Negative Emissions," *Science 354*, no. 6309 (October 14, 2016): 182–83, https://doi.org/10.1126/science.aah4567; see also Kevin Anderson, "Duality in Climate Science," *Nature Geoscience 8*, no. 12 (December 2015): 898–900, https://doi.org/10.1038/ngeo2559; Oliver Geden, "Policy: Climate Advisers Must Maintain Integrity," *Nature 521*, no. 7550 (May 2015): 27–28, https://doi.org/10.1038/521027a; Oliver Geden, "The Paris Agreement and the Inherent Inconsistency of Climate Policymaking," *WIREs Climate Change 7*, no. 6 (November 2016): 790–97, https://doi.org/10.1002/wcc.427; Henry Shue, "Climate Dreaming: Negative Emissions, Risk Transfer, and Irreversibility," *SSRN Electronic Journal*, 2017, https://doi.org/10.2139/ssrn.2940987; Alice Larkin et al., "What If Negative Emission Technologies Fail at Scale? Implications of the Paris Agreement for Big Emitting Nations," *Climate Policy 18*, no. 6 (July 3, 2018): 690–714, https://doi.org/10.1080/14693062.2017.1346498; Kate Dooley and Sivan Kartha, "Land-Based Negative Emissions: Risks for Climate Mitigation and Impacts on Sustainable Development," *International Environmental Agreements: Politics, Law and Economics 18*, no. 1 (February 2018): 79–98, https://doi.org/10.1007/s10784-017-9382-9; Kate Dooley, Peter Christoff, and Kimberly A. Nicholas, "Co-Producing Climate Policy and Negative Emissions: Trade-Offs for Sustainable Land-Use," *Global Sustainability 1* (2018): e3, https://doi.org/10.1017/sus.2018.6; Minx et al., "Negative Emissions—Part 1," 20; Duncan McLaren and Nils Markusson, "The Co-Evolution of Technological Promises, Modelling, Policies and Climate Change Targets," *Nature Climate Change 10*, no. 5 (May 2020): 392–97, https://doi.org/10.1038/s41558-020-0740-1; Wim Carton, "Carbon Unicorns and Fossil Futures: Whose Emission Reduction Pathways Is the IPCC Performing?," in *Has It Come to This? The Promises and Perils of Geoengineering on the Brink*, ed. J. P. Sapinski, Holly Jean Buck, and Andreas Malm, *Nature, Society, and Culture* (New Brunswick, NJ: Rutgers University Press, 2020).
84. David Ho, "Carbon Dioxide Removal Is not a Current Climate Solution—We Need to Change the Narrative," *Nature* 616, no. 7955 (2023): 9, https://doi.org/10.1038/d41586-023-00953-x.
85. Rogelj et al., "A New Scenario Logic for the Paris Agreement Long-Term Temperature Goal."
86. Zeke Hausfather, "Analysis: When Might the World Exceed 1.5C and 2C of Global Warming?," *Carbon Brief*, December 4, 2020, https://

www.carbonbrief.org/analysis-when-might-the-world-exceed-1-5c-and-2c-of-global-warming/.

CHAPTER 6

1. Intergovernmental Panel on Climate Change, "Climate Change 2022: Impacts, Adaptation, and Vulnerability," 2022, 7, https://report.ipcc.ch/ar6/wg2/IPCC_AR6_WGII_FullReport.pdf.
2. UNFCCC, "Paris Agreement (All Language Versions)," December 12, 2015, https://unfccc.int/process/conferences/pastconferences/paris-climate-change-conference-november-2015/paris-agreement.
3. "What You Need to Know about the Climate Change Resilience Rating System," *World Bank,* January 25, 2021, https://www.worldbank.org/en/news/feature/2021/01/25/what-you-need-to-know-about-the-climate-change-resilience-rating-system.
4. "National Climate Task Force," *The White House,* accessed August 31, 2023, https://www.whitehouse.gov/climate/.
5. Jason Stanley, *How Propaganda Works* (Princeton, NJ: Princeton University Press, 2017), 43.
6. C. S. Holling, "Resilience and Stability of Ecological Systems." *Annual Review of Ecology and Systematics 4* (1973): 1–23, 14, http://www.jstor.org/stable/2096802.
7. National Academies of Sciences, Engineering, and Medicine, *Disaster Resilience: A National Imperative* (Washington, D.C.: National Academies Press, 2012), 16, https://doi.org/10.17226/13457.
8. Jakob Zscheischler et al., "Future Climate Risk from Compound Events," *Nature Climate Change 8*, no. 6 (June 2018): 469–77, https://doi.org/10.1038/s41558-018-0156-3; Colin Raymond et al., "Increasing Spatiotemporal Proximity of Heat and Precipitation Extremes in a Warming World Quantified by a Large Model Ensemble," *Environmental Research Letters 17*, no. 3 (March 1, 2022): 035005, https://doi.org/10.1088/1748-9326/ac5712; Kumar P. Tripathy et al., "Climate Change Will Accelerate the High-End Risk of Compound Drought and Heatwave Events," *Proceedings of the National Academy of Sciences 120*, no. 28 (July 11, 2023): e2219825120, https://doi.org/10.1073/pnas.2219825120.
9. Riazat Butt, "After Devastating Floods in Pakistan, Some Have Recovered but Many Are Struggling a Year Later," *AP News*, June 22, 2023, sec. Climate, https://apnews.com/article/pakistan-flood-anniversary-ebd91932d0452d47c3b0c4bd2a656f38.

10. Christopher Flavelle and Bryan Tarnowski, "As the Great Salt Lake Dries Up, Utah Faces an 'Environmental Nuclear Bomb,'" *New York Times*, June 7, 2022, sec. Climate, https://www.nytimes.com/2022/06/07/climate/salt-lake-city-climate-disaster.html.
11. Matthew Rozsa, "Climate Change Will Raise Sea Levels, Cause Apocalyptic Floods and Displace Almost a Billion People," *Salon*, August 28, 2023, https://www.salon.com/2023/08/28/climate-change-will-raise-sea-levels-cause-apocalyptic-floods-and-displace-almost-a-bill ion-people/; Kendra Pierre-Louis, "How Rising Groundwater Caused by Climate Change Could Devastate Coastal Communities," *MIT Technology Review*, accessed August 31, 2023, https://www.technolog yreview.com/2021/12/13/1041309/climate-change-rising-groundwa ter-flooding/.
12. Intergovernmental Panel on Climate Change, "Climate Change 2022: Impacts, Adaptation, and Vulnerability," 26.
13. Michael E. Mann, *The New Climate War: The Fight to Take Back Our Planet*, First edition (New York: PublicAffairs, 2021), 177.
14. Danny MacKinnon and Kate Driscoll Derickson, "From Resilience to Resourcefulness: A Critique of Resilience Policy and Activism," *Progress in Human Geography 37*, no. 2 (April 2013): 253–70, https://doi.org/10.1177/0309132512454775.
15. Ashley Dawson, *Extreme Cities: The Peril and Promise of Urban Life in the Age of Climate Change* (London: Verso, 2017), 157.
16. Benjamin Storrow, "Batteries Keep Texas Grid Humming in Heat Waves," *E&E News*, August 22, 2023, https://subscriber.politicopro.com/article/eenews/2023/08/22/batteries-keep-texas-grid-humm ing-in-heat-waves-00112136; Arpan Varghese and Scott Disavino, "Wind, Solar Help Texas Meet Record Power Demand during Heat Wave," *Reuters*, June 30, 2023, sec. Environment, https://www.reut ers.com/business/environment/wind-solar-help-texas-meet-record-power-demand-during-heat-wave-2023-06-30/; Catherine Rampell, "Renewables Are Saving Texas. Again. So Give Them Their Due," *Washington Post*, July 4, 2023, https://www.washingtonpost.com/opini ons/2023/07/04/renewable-energy-texas-heat/.
17. Eric Larson et al., "Net-Zero America: Potential Pathways, Infrastructure, and Impacts" (Princeton, NJ: Princeton University, December 15, 2020); Auke Hoekstra, "How a Recent Attack on 100% Renewable Models Might End the Trench Warfare and Benefit Our Climate," *NEON Research (blog),* July 31, 2023, https://neonresearch.nl/thebti-rebuttal/; Mark Z. Jacobson, *No Miracles Needed: How Today's Technology Can Save*

Our Climate and Clean Our Air (Cambridge, UK: Cambridge University Press, 2023).

18. Marco Rubio, "America Deserves Better than Democrats' Fake Green Agenda," *Washington Times*, April 18, 2023, https://www.washingtontimes.com/news/2023/apr/18/america-deserves-better-than-democrats-fake-green-/.
19. Laura Cassels, "Avoiding Words 'Climate Change,' DeSantis Says Global-Warming Concerns Involve 'Left-Wing Stuff,'" *Florida Phoenix*, December 7, 2021, https://floridaphoenix.com/2021/12/07/avoiding-words-climate-change-desantis-says-global-warming-concerns-involve-left-wing-stuff/.
20. "Resilient Florida Program," Florida Department of Environmental Protection, accessed September 1, 2023, https://floridadep.gov/ResilientFlorida; Alex Harris, "Florida Gets Another $404 Million for Climate Change Prep. It Needs Billions More," *WUSF Public Media*, February 2, 2022, sec. Environment, https://wusfnews.wusf.usf.edu/environment/2022-02-02/florida-gets-another-404-million-for-climate-change-prep-it-needs-billions-more.
21. Craig Pittman, "Hurricane Ian Proved Why Ron DeSantis's Version of Climate Resilience Is a Disaster," *New York Times*, October 13, 2022, sec. Opinion, https://www.nytimes.com/2022/10/13/opinion/environment/ron-desantis-hurricane-ian-climate-change.html.
22. "Governor Ron DeSantis Announces Award of Nearly $20 Million for 98 Projects through the Resilient Florida Grant Program," *Florida Commerce*, May 3, 2022, https://floridajobs.org/news-center/DEO-Press/deo-press-2022/2022/05/03/governor-ron-desantis-announces-award-of-nearly-$20-million-for-98-projects-through-the-resilient-florida-grant-program; "Governor Ron DeSantis Announces Award of More than $275 Million through the Resilient Florida Grant Program," Ron DeSantis, 46th Governor of Florida, February 6, 2023, https://www.flgov.com/2023/02/06/governor-ron-desantis-announces-award-of-more-than-275-million-through-the-resilient-florida-grant-program/; Brandon Hogan, "Gov. DeSantis Announces $300M for Flood Resilience Projects, $11.1M in Central Florida," WKMG, July 14, 2023, https://www.clickorlando.com/news/local/2023/07/14/gov-desantis-announces-300m-for-flood-resilience-projects-111m-in-central-florida/; "The Nature Conservancy in Florida 2022 Florida Legislative Results Report" (The Nature Conservancy, n.d.), https://www.nature.org/content/dam/tnc/nature/en/documents/TNC-Florida-2022-Legislative-Session-Final-Report-TNC.pdf.

23. "An Act Relating to Preemption over Restriction of Utility Services; Creating s. 366.032, F.S." (*Florida House of Representatives,* 2021), https://www.flsenate.gov/Session/Bill/2021/919/BillText/er/PDF.
24. State Board of Administration of Florida, "A Resolution Directing an Update to the Investment Policy Statement and Proxy Voting Policies for the Florida Retirement System Defined Benefit Pension Plan," August 23, 2022, https://www.flgov.com/wp-content/uploads/2022/08/ESG-Resolution-Final.pdf.
25. Mark Brnovich et al., "Dear Mr. Fink," August 4, 2022, https://www.texasattorneygeneral.gov/sites/default/files/images/executive-management/BlackRock%20Letter.pdf.
26. Dana Drugmand, "Public Pension Funds Have Lost Billions on Their Fossil Fuel Investments: New Analysis," *DeSmog*, June 28, 2023, https://www.desmog.com/2023/06/27/public-pension-funds-lost-billions-on-their-fossil-fuel-investments/.
27. UN Environment Programme, "Emissions Gap Report 2022," October 21, 2022, XIVV, http://www.unep.org/resources/emissions-gap-report-2022.
28. Intergovernmental Panel on Climate Change, "Sixth Assessment Report, Working Group II: Impacts, Adaptation, and Vulnerability. Overarching Frequently Asked Questions and Answers," June 2023, https://www.ipcc.ch/report/ar6/wg2/downloads/faqs/IPCC_AR6_WGII_Overaching_OutreachFAQ6.pdf.
29. Shalanda Baker, "Anti-Resilience: A Roadmap for Transformational Justice within the Energy System," *Harvard Civil Rights-Civil Liberties Law Review* 54 (2019): 6.
30. "DOE Justice40 Covered Programs," Energy.gov, accessed November 2, 2023, https://www.energy.gov/justice/doe-justice40-covered-programs.
31. White House Environmental Justice Advisory Council, "Justice40: Climate and Economic Justice Screening Tool & Executive Order 12898 Revisions Interim Final Recommendations" (Washington, D.C., May 13, 2021).
32. Rhiana Gunn-Wright, "Our Green Transition May Leave Black People Behind," *Hammer & Hope*, Summer 2023, https://hammerandhope.org/article/climate-green-new-deal.
33. "War Production," The War | Ken Burns | PBS, accessed September 1, 2023, https://www.pbs.org/kenburns/the-war/war-production.
34. Gunn-Wright, "Our Green Transition May Leave Black People Behind."

35. Erica Chenoweth and Maria J. Stephan, *Why Civil Resistance Works: The Strategic Logic of Nonviolent Conflict*, Columbia Studies in Terrorism and Irregular Warfare (New York: Columbia University Press, 2011).
36. Gallup, "Most Important Problem," Gallup.com, February 2023, https://news.gallup.com/poll/1675/Most-Important-Problem.aspx.
37. Domenico Montanaro, "Three-Quarters of Republicans Prioritize the Economy over Climate Change," *NPR*, August 3, 2023, sec. Politics, https://www.npr.org/2023/08/03/1191678009/climate-change-republicans-economy-natural-disasters-biden-trump-poll.
38. Lecture at Southern Illinois University Carbondale (February 13, 2014).
39. Rebecca Solnit, *Hope in the Dark: Untold Histories, Wild Possibilities*, Third edition (Chicago, IL: Haymarket Books, 2016).

AFTER WORDS: WALKING THE TALK

1. Yale Program on Climate Change Communication, "Yale Climate Opinion Maps 2021," *Yale Program on Climate Change Communication (blog),* accessed November 1, 2023, https://climatecommunication.yale.edu/visualizations-data/ycom-us/.

Index

For the benefit of digital users, indexed terms that span two pages (e.g., 52–53) may, on occasion, appear on only one of those pages.